NEWTONIAN ATTRACTION

NEWTONIAN ATTRACTION

A. S. RAMSEY

*Rationem vero harum Gravitatis proprietatum ex Phœnomenis
nondum potui deducere, & Hypotheses non fingo.*
NEWTON

CAMBRIDGE UNIVERSITY PRESS

CAMBRIDGE
LONDON   NEW YORK   NEW ROCHELLE
MELBOURNE   SYDNEY

Published by the Press Syndicate of the University of Cambridge
The Pitt Building, Trumpington Street, Cambridge CB2 1RP
32 East 57th Street, New York, NY 10022, USA
296 Beaconsfield Parade, Middle Park, Melbourne 3206, Australia

First published as *An introduction to the theory of Newtonian
attraction* 1940
Reprinted 1949, 1952, 1956, 1959, 1961
Reissued as *Newtonian Attraction* 1981

Printed in Great Britain at the University Press, Cambridge

British Library Cataloguing in Publication Data

Ramsey, A. S.
[An introduction to the theory of Newtonian attraction.]
Newtonian attraction

1. Mechanics
I. Title II. Newtonian attraction
531   QA805

ISBN 0 521 09193 4

PREFACE

This book has been written at the suggestion of some students who have been unable to find a book on the subject suitable for their requirements, and also with a view to completing the series of text-books on Mechanics which I have produced in recent years.

The book is only an introduction to a subject which, in the eighteenth and nineteenth centuries, occupied the interest and attention of many eminent mathematicians. Nowadays the theory of potential is usually studied in connection with electricity and magnetism, which is given priority as a subject of study because of its wider developments and importance in relation to everyday affairs, so that there is a tendency to overlook the theory of the attraction of gravitating solids or regard it as a subject to be 'crowded out'. But the subject is of importance in connection with dynamical astronomy and the Figure of the Earth, and Todhunter's history of the subject 'from Newton to Laplace' ran to nearly a thousand pages.

In the early chapters I have not forgotten the needs of students reading for a pass degree, though the book as a whole is intended for an honours course. A good many of the theorems in pure mathematics specially required for the subject are collected in the first chapter, but students who prefer to do so may begin to read at Chapter II and only refer back to Chapter I as the need arises. The later chapters are on Green's theorem, Harmonic functions and the Attraction of ellipsoids.

I have indicated the source of the examples when taken from examination papers, the abbreviations M. T., C., and P. denoting the Mathematical Tripos, a College or Intercollegiate Examination, and the Preliminary Examination in Mathematics in the University of Cambridge, and I gratefully acknowledge the kind permission of the Senate of the University of London to use questions from the examinations of that University, the source of which is also indicated.

A line in each set of examples indicates a rough division of those which are easier from those which are more difficult of solution.

The proofs have been read by my friend Dr S. Verblunsky and again I tender to him my thanks. The book has been greatly improved by his valuable criticism and useful suggestions. I also wish to thank the staff of the University Press for their careful composition and assistance in eliminating errors.

A. S. R.

CAMBRIDGE
May 1940

CONTENTS

Chapter I: PRELIMINARY MATHEMATICS

Chapter II: GRAVITATIONAL ATTRACTION AND POTENTIAL. SIMPLE APPLICATIONS

Chapter III: ATTRACTION AND POTENTIAL AT INTERNAL POINTS. SPHERES

Chapter VII: ATTRACTION OF ELLIPSOIDS

Chapter I

PRELIMINARY MATHEMATICS

1·1. We propose in this chapter to give a brief account of some mathematical ideas and propositions which the reader will find of use in later chapters.

1·2. Surface and volume integrals. In the theory of attractions it is necessary to make considerable use of surface and volume integrals, and the applications of the theory frequently involve the evaluation of such integrals. The reader is therefore advised to acquire some familiarity with methods of integration from a textbook on the subject. We propose here merely to explain what is implied when such symbols as

$$\int f(x,y,z)\,dS \quad \text{and} \quad \int f(x,y,z)\,dv$$

are used to denote integration over a surface or through a volume, and then to prove some important propositions connecting surface and volume integrals.

A definite integral of a function of one variable, say $\int_a^b f(x)\,dx$, may be defined thus: let the interval from a to b on the x-axis be divided into any number of sub-intervals $\delta_1, \delta_2, \ldots \delta_n$, and let f_r denote the value of $f(x)$ at some point on δ_r. Then, provided that the limit as $n \to \infty$ of $\sum_{r=1}^{n} f_r \delta_r$ exists and is independent of the method of division into sub-intervals and of the choice of the point on δ_r at which the value of $f(x)$ is taken, this limit is the definite integral of $f(x)$ from a to b. It can be shewn that the conditions for the existence of the limit are satisfied if $f(x)$ is a continuous function.

In the same way we may define $\int f(x,y,z)\,dS$ over a given surface; let the given surface be divided into any number of small parts $\delta_1, \delta_2, \dots \delta_n$ and let f_r denote the value of $f(x, y, z)$ at some point on δ_r, then the limit as $n \to \infty$ of $\sum\limits_{r=1}^{n} f_r \delta_r$, provided the limit exists under the same conditions as aforesaid, is defined to be the integral $\int f(x, y, z) \, dS$ over the given surface.

Any difficulty as to the precise meaning to be attributed to 'area of a curved surface' may be avoided thus: after choosing the point on each sub-division δ_r of the surface at which the value of $f(x, y, z)$ is taken, project this element of surface on to the tangent plane at the chosen point, and take the plane projection of the element as the measure of δ_r in forming the sum.

The integral $\int f(x, y, z) \, dv$ through a given volume may be defined in a similar way.

The evaluation of a surface integral of course involves a double integration, and that of a volume integral a triple integration, and when it is desired to exhibit the forms chosen for the elements of area or elements of volume such symbols as $\iint f(x, y, z) \, dx \, dy$ and $\iiint f(x, y, z) \, dx \, dy \, dz$ are used; but when there is no doubt as to what is implied a single integration symbol $\int$ is employed as above for the representation of an integral along a line or over a surface or throughout a volume.

1·3. Green's Theorem. *If u, v, w are functions of x, y, z which have continuous derivatives with respect to x, y, z respectively throughout a singly-connected region bounded by a closed surface S free from singularities and l, m, n denote the direction cosines of the normal drawn outwards from a point on an element dS of the surface, then*

$$\iiint \left(\frac{\partial u}{\partial x} + \frac{\partial v}{\partial y} + \frac{\partial w}{\partial z} \right) dx \, dy \, dz = \iint (lu + mv + nw) \, dS,$$

where the surface integral is taken over the boundary and the volume integral throughout the region enclosed.

Consider the integral $\iiint \dfrac{\partial w}{\partial z}\,dx\,dy\,dz$. Take a prism of cross-

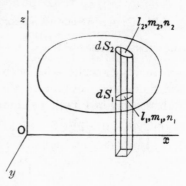

section $dx\,dy$ parallel to the axis of z. Let it intersect the surface S in elements dS_1, dS_2 and let normals to these elements have direction cosines l_1, m_1, n_1; l_2, m_2, n_2. Since the elements dS_1, dS_2 project into an area $dx\,dy$ on the xy-plane, taking account of signs we have, from the figure,

$$dx\,dy = -n_1 dS_1 = n_2 dS_2 \quad \ldots\ldots\ldots\ldots(1).$$

Hence, by integration with regard to z, we have for this particular prism

$$\int \frac{\partial w}{\partial z}\,dx\,dy\,dz = (w\,dx\,dy)_2 - (w\,dx\,dy)_1,$$

the limiting values corresponding to the intersections of the prism with the surface, so that

$$\int \frac{\partial w}{\partial z}\,dx\,dy\,dz = n_2 w_2\,dS_2 + n_1 w_1\,dS_1.$$

The integration through the whole bounded region is now completed by summing for all similar parallel prisms which intersect the surface S, so that

$$\iiint \frac{\partial w}{\partial z}\,dx\,dy\,dz = \iint nw\,dS.$$

The integrals relating to the functions u and v may be treated similarly by considering prisms parallel to the x and y axes, so that

$$\iiint \left(\frac{\partial u}{\partial x} + \frac{\partial v}{\partial y} + \frac{\partial w}{\partial z}\right) dx\,dy\,dz = \iint (lu + mv + nw)\,dS \quad \ldots(2).$$

This is **Green's Theorem*** connecting a volume integral with a surface integral over the boundary of the volume.

1·31. Green's Second Theorem. In the last theorem make the substitutions $u = V \dfrac{\partial V'}{\partial x}$, $v = V \dfrac{\partial V'}{\partial y}$, $w = V \dfrac{\partial V'}{\partial z}$, where V, V' are functions which with their first and second derivatives are finite and continuous through the region considered. Substituting in **1·3** (2), we obtain

$$\iiint \left(\frac{\partial V}{\partial x}\frac{\partial V'}{\partial x} + \frac{\partial V}{\partial y}\frac{\partial V'}{\partial y} + \frac{\partial V}{\partial z}\frac{\partial V'}{\partial z} \right) dx\, dy\, dz$$

$$+ \iiint V \left(\frac{\partial^2 V'}{\partial x^2} + \frac{\partial^2 V'}{\partial y^2} + \frac{\partial^2 V'}{\partial z^2} \right) dx\, dy\, dz$$

$$= \iint V \left(l\frac{\partial V'}{\partial x} + m\frac{\partial V'}{\partial y} + n\frac{\partial V'}{\partial z} \right) dS \quad \ldots\ldots\ldots(1).$$

But if $\partial/\partial n_1$ denotes differentiation in the direction of the *outward-drawn* normal to dS, we have $l, m, n \equiv \dfrac{\partial x}{\partial n_1}, \dfrac{\partial y}{\partial n_1}, \dfrac{\partial z}{\partial n_1}$, so that

$$l\frac{\partial V'}{\partial x} + m\frac{\partial V'}{\partial y} + n\frac{\partial V'}{\partial z} = \frac{\partial V'}{\partial x}\frac{\partial x}{\partial n_1} + \frac{\partial V'}{\partial y}\frac{\partial y}{\partial n_1} + \frac{\partial V'}{\partial z}\frac{\partial z}{\partial n_1}$$

$$= \frac{\partial V'}{\partial n_1}.$$

Hence (1) is equivalent to

$$\iiint \left(\frac{\partial V}{\partial x}\frac{\partial V'}{\partial x} + \frac{\partial V}{\partial y}\frac{\partial V'}{\partial y} + \frac{\partial V}{\partial z}\frac{\partial V'}{\partial z} \right) dx\, dy\, dz$$

$$= \iint V \frac{\partial V'}{\partial n_1} dS - \iiint V \left(\frac{\partial^2 V'}{\partial x^2} + \frac{\partial^2 V'}{\partial y^2} + \frac{\partial^2 V'}{\partial z^2} \right) dx\, dy\, dz,$$

and since V and V' are clearly interchangeable, this also

$$= \iint V' \frac{\partial V}{\partial n_1} dS - \iiint V' \left(\frac{\partial^2 V}{\partial x^2} + \frac{\partial^2 V}{\partial y^2} + \frac{\partial^2 V}{\partial z^2} \right) dx\, dy\, dz \quad \ldots(2),$$

---

* George Green (1793–1841), published his *Essay on the Application of Mathematical Analysis to the Theory of Electricity and Magnetism* in 1828. At the age of 40 he came to Cambridge and was fourth wrangler in 1837.

where the volume integrals are taken throughout the region considered and the surface integrals over its boundary.

1·4. Solid angles. The solid angle of a cone *is measured by the area intercepted by the cone on the surface of a sphere of unit radius having its centre at the vertex of the cone.*

The solid angle subtended at a point by a surface of any form is measured by the solid angle of the cone whose vertex is at the given point and whose base is the given surface.

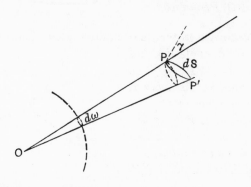

Let PP' be a small element of area dS which subtends a solid angle $d\omega$ at O.

Let the normal to dS make an acute angle γ with OP, and let $OP = r$. Then the cross-section at P of the cone which dS subtends at O is $dS \cos \gamma$, and this cross-section and the small area $d\omega$ intercepted on the unit sphere are similar figures, so that

$$dS \cos \gamma : d\omega = r^2 : 1.$$

Whence $\qquad d\omega = (dS \cos \gamma)/r^2$

or $\qquad\qquad dS = r^2 \sec \gamma \, d\omega \qquad\Bigg\}\qquad$ ....(1).

It follows that the area of a finite surface can be represented as an integral over a spherical surface, thus

$$S = \int r^2 \sec \gamma \, d\omega \qquad ....(2),$$

with suitable limits of integration.

1·41. $d\omega$ in polar co-ordinates. $d\omega$ is an element of the
surface of a unit sphere. Let the
element be $PQRS$ bounded by
meridians and small circles, where
the angular co-ordinates of P are
θ, ϕ. Then since the arc PS subtends an angle $d\phi$ at the centre of
a circle of radius $\sin\theta$, therefore
$PS = \sin\theta\,d\phi$; and $PQ = d\theta$, so that

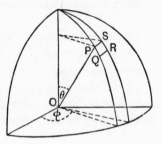

$$d\omega = PQ . PS = \sin\theta\,d\theta\,d\phi.$$

1·42. Volumes in terms of solid angles. Consider a cone
of small solid angle $d\omega$ and
length R. Let PP', QQ' be
cross-sections at distances r,
$r + dr$ from the vertex O.

Then the area $PP' = r^2 d\omega$,
and, neglecting an infinitesimal of higher order, the
volume of the frustum $PQQ'P'$
is $r^2 d\omega\,dr$.

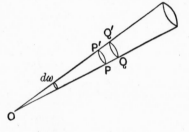

Hence the volume of the cone $= d\omega \displaystyle\int_0^R r^2 dr$

$$= \tfrac{1}{3}R^3 d\omega.$$

Hence if O be an origin inside a region bounded by a continuous surface such that lines from O each meet the surface
in one point only and we divide up the region into narrow
cones with vertices at O and R denotes the radius vector or
length of the cone of solid angle $d\omega$, then

$$\frac{1}{3}\int R^3 d\omega$$

will represent the volume of the region.

1·5. Scalar functions of position and their gradients.
Let $V(x, y, z)$ be a single-valued continuous function of the
position of a point in some region of space. Suppose that the

function V is not constant throughout any region, so that the equation

$$V(x, y, z) = \text{const.}$$

represents a surface. We assume that through each point of the region in which V is defined, there passes a surface $V = \text{const.}$ We also assume that at every point P on this surface there is a definite normal PN and that

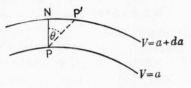

the tangent plane at P varies continuously with the position of P on the surface.

It follows from the definition of V that two surfaces

$$V(x, y, z) = a \quad \text{and} \quad V(x, y, z) = b$$

cannot intersect; for if they had a common point it would be a point at which V had more than one value, in contradiction to the hypothesis that V is a single-valued function.

Consider two neighbouring surfaces

$$V = a \quad \text{and} \quad V = a + da.$$

Let P, P' be points on each and let the normal at P to $V = a$ meet $V = a + da$ in N. For small values of da, PN will also be a normal to $V = a + da$.

Then using V_P to denote the value of V at P, we have

$$\frac{V_{P'} - V_P}{PP'} = \frac{da}{PP'} = \frac{V_N - V_P}{PP'} = \frac{V_N - V_P}{PN} \cdot \frac{PN}{PP'}$$

$$= \frac{V_N - V_P}{PN} \cos \theta,$$

where θ is the angle NPP'.

Now if $PP' = ds$ and $PN = dn$, and we make da and therefore also ds and dn tend to zero, the limit of $(V_{P'} - V_P)/PP'$ is the rate of increase of V in the direction ds and is denoted by $\dfrac{\partial V}{\partial s}$; and similarly the limit of $(V_N - V_P)/PN$ is the rate of increase of V in the normal direction dn and is denoted by $\dfrac{\partial V}{\partial n}$, and we have

$$\frac{\partial V}{\partial s} = \frac{\partial V}{\partial n} \cos \theta \quad \dots\dots\dots\dots\dots\dots(1).$$

Thus we have proved that the space rate of increase of V in any direction ds is the component in that direction of its space rate of increase in the direction normal to the surface $V = \text{const.}$; or that if we construct a vector of magnitude $\partial V/\partial n$ in direction PN, then the component of this vector in any direction is the space rate of increase of V in that direction.

This vector is called the **gradient** of V and written **grad** V.

To recapitulate: V is a continuous *scalar* function of position having a definite single value at each point of a certain region of space, and the **gradient** of V is defined in this way: through any point P in the region there passes a surface $V = \text{const.}$, then a *vector*, normal to this surface at P, whose magnitude is the space rate of increase of V in this normal direction, is defined to be the *gradient of V at P*, and it has the property that its component in any direction gives the space rate of increase of V in that direction. It is clear that the gradient measures the greatest rate of increase of V at a point.

1·6. Vectorial methods. We shall denote vectors by letters in Clarendon type. The magnitude of $\mathbf{P}$ is denoted by $|\mathbf{P}|$. We recall that vectors are compounded by the parallelogram law and that they obey the commutative and associative laws of addition, viz. that

$$\mathbf{P} + \mathbf{Q} = \mathbf{Q} + \mathbf{P}, \text{ and that } \mathbf{P} + (\mathbf{Q} + \mathbf{R}) = (\mathbf{P} + \mathbf{Q}) + \mathbf{R}.$$

The product of a vector $\mathbf{P}$ and a scalar m is a vector whose magnitude is the product of the magnitude of $\mathbf{P}$ and m, i.e. $m|\mathbf{P}|$, whose direction is that of $\mathbf{P}$ or the opposite according as m is positive or negative. Multiplication by scalars follows the ordinary rules of algebra, viz.

$$m(n\mathbf{P}) = n(m\mathbf{P}) = (mn)\mathbf{P};$$
$$m(\mathbf{P} + \mathbf{Q}) = m\mathbf{P} + m\mathbf{Q}$$

and $\qquad\qquad (m+n)\mathbf{P} = m\mathbf{P} + n\mathbf{P}.$

A vector **s** is called a unit vector when $|\,\mathbf{s}\,| = 1$. Any vector whose direction is that of a unit vector **s** can be represented by $m\mathbf{s}$, where m is a scalar.

For convenience in representing vectors by rectangular components we introduce three **fundamental unit vectors** **i**, **j**, **k** whose directions are mutually perpendicular; these directions being those of the co-ordinate axes Ox, Oy, Oz. Then, if P_x, P_y, P_z be the projections of the vector **P** on the axes, we have a vector equation

$$\mathbf{P} = P_x\mathbf{i} + P_y\mathbf{j} + P_z\mathbf{k} \quad \ldots\ldots\ldots\ldots(1).$$

1·61. Scalar product. Let **P**, **Q** be two vectors and ϕ the angle between their positive directions, then the expression

$$|\,\mathbf{P}\,|\,|\,\mathbf{Q}\,|\cos\phi$$

is called the **scalar product** of the vectors and is denoted by **(PQ)** or **(P, Q)**, or simply **PQ**.

The scalar product of two vectors at right angles is zero, so that for the fundamental vectors **i**, **j**, **k** we have

$$\mathbf{ij} = \mathbf{jk} = \mathbf{ki} = 0 \quad \text{and} \quad \mathbf{ii} = \mathbf{jj} = \mathbf{kk} = 1.$$

Hence, from **1·6** (1),

$$\mathbf{PQ} = (P_x\mathbf{i} + P_y\mathbf{j} + P_z\mathbf{k})(Q_x\mathbf{i} + Q_y\mathbf{j} + Q_z\mathbf{k});$$

and on multiplying out this gives

$$\mathbf{PQ} = P_xQ_x + P_yQ_y + P_zQ_z \quad \ldots\ldots\ldots\ldots(1).$$

1·62. Again if V denotes a function of x, y, z, it follows from **1·5** that the projections on the axes of the gradient of V are $\partial V/\partial x$, $\partial V/\partial y$, $\partial V/\partial z$. Hence we have the relation

$$\mathbf{grad}\,V = \mathbf{i}\frac{\partial V}{\partial x} + \mathbf{j}\frac{\partial V}{\partial y} + \mathbf{k}\frac{\partial V}{\partial z},$$

and the equivalence of the operators **grad** and $\mathbf{i}\dfrac{\partial}{\partial x} + \mathbf{j}\dfrac{\partial}{\partial y} + \mathbf{k}\dfrac{\partial}{\partial z}$, when the latter is applied to a scalar function. This latter operator is known as the **Hamiltonian operator*** and denoted by ∇. When applied to a *scalar* function it turns it into a vector.

---

* After Sir William Rowan Hamilton (1805–1865), Astronomer Royal for Ireland, greatest of Irish Mathematicians, inventor of *Quaternions.*

We shall see in **1·65** that when ∇ is applied to a vector it turns it into a scalar.

The reader who wishes for further information on the subject of vector algebra may find it in the author's *Dynamics*, Part II, Appendix.

1·63. A vector field. If to every point of a given region there corresponds a definite vector **A**, which may vary in magnitude and direction from point to point, then the region is called a **vector field**, or the field of the vector **A**; e.g. gravitational field, electric field.

1·64. Flux of a vector. If a surface S be drawn in the field of a vector **A** and A_n denotes the component of **A** normal to an element dS of the surface, then the integral $\int A_n dS$ is called **the flux of A through** S. Since a surface has two sides the sense of the normal must be taken into account, and the sign of the flux is changed when the sense of the normal is changed. The flux of a vector through a surface is clearly a scalar magnitude, in fact if we use the symbol **dS** to represent a vector of magnitude dS directed along the normal to the surface, then the scalar product of the vectors **A** and **dS** denoted by (**A dS**) is the same thing as the subject of integration $A_n dS$, so that the flux of the vector may also be represented by $\int (\mathbf{A\,dS})$.

1·65. Divergence of a vector field. Let **A** denote a vector field which has no discontinuities throughout a given region. Let δv denote any small element of volume containing a point P in the region and let $\int A_n dS$ denote the outward flux of **A** through the boundary of δv, then the limit as $\delta v \to 0$ of $\left(\int A_n dS \right) \Big/ \delta v$ is defined to be the divergence of **A** at the point P and denoted by div **A**.

Let P be the point (x, y, z). Let A_x, A_y, A_z denote rectangular components of the vector **A**.

Let l, m, n be direction cosines of the normal to dS.

Then
$$A_n = lA_x + mA_y + nA_z,$$

so that
$$\int A_n\, dS = \int (lA_x + mA_y + nA_z)\, dS$$

$$= \int \left(\frac{\partial A_x}{\partial x} + \frac{\partial A_y}{\partial y} + \frac{\partial A_z}{\partial z} \right) dv,$$

by Green's theorem (**1·3**), where the volume integral is taken through δv. The value of the integral is the product of δv and the mean value of the integrand, say the value at a point Q. As $\delta v \to 0$, this mean value tends to the value at P.

Hence
$$\operatorname{div} \mathbf{A} = \lim_{\delta v \to 0} \frac{\displaystyle \int A_n\, dS}{\delta v} = \frac{\partial A_x}{\partial x} + \frac{\partial A_y}{\partial y} + \frac{\partial A_z}{\partial z} \quad \ldots\ldots(1),$$

and we note that this result is independent of the form of the element of volume δv which surrounds the point P, and also independent of the way in which the element contracts to zero.

Further, in the notation of **1·61, 1·62**, we have

$$\nabla \mathbf{A} = \left(\mathbf{i} \frac{\partial}{\partial x} + \mathbf{j} \frac{\partial}{\partial y} + \mathbf{k} \frac{\partial}{\partial z} \right) (\mathbf{i} A_x + \mathbf{j} A_y + \mathbf{k} A_z)$$

$$= \frac{\partial A_x}{\partial x} + \frac{\partial A_y}{\partial y} + \frac{\partial A_z}{\partial z}$$

$$= \operatorname{div} \mathbf{A},$$

so that when the operation ∇ is performed on a vector $\mathbf{A}$ the result is the scalar function $\operatorname{div} \mathbf{A}$.

Also if $\mathbf{A} = \operatorname{grad} V$ (**1·5**), where V is a scalar function of x, y, z, so that

$$A_x,\ A_y,\ A_z \equiv \frac{\partial V}{\partial x},\quad \frac{\partial V}{\partial y},\quad \frac{\partial V}{\partial z},$$

we have
$$\operatorname{div} \operatorname{grad} V = \operatorname{div} \mathbf{A}$$

$$= \frac{\partial A_x}{\partial x} + \frac{\partial A_y}{\partial y} + \frac{\partial A_z}{\partial z}$$

$$= \frac{\partial^2 V}{\partial x^2} + \frac{\partial^2 V}{\partial y^2} + \frac{\partial^2 V}{\partial z^2} \quad \ldots\ldots\ldots\ldots(2).$$

The operator $\dfrac{\partial^2}{\partial x^2} + \dfrac{\partial^2}{\partial y^2} + \dfrac{\partial^2}{\partial z^2}$ is usually denoted by ∇^2, for it is the same as

$$\left(\mathbf{i}\,\frac{\partial}{\partial x} + \mathbf{j}\,\frac{\partial}{\partial y} + \mathbf{k}\,\frac{\partial}{\partial z}\right)\left(\mathbf{i}\,\frac{\partial}{\partial x} + \mathbf{j}\,\frac{\partial}{\partial y} + \mathbf{k}\,\frac{\partial}{\partial z}\right).$$

The reader will note that the operator ∇ may be applied either to a scalar, turning it into a vector, or to a vector, turning it into a scalar; i.e. that ∇ means '**grad**' when applied to a scalar and it means 'div' when applied to a vector. But the operator '**grad**' only has a meaning when applied to a scalar function, and 'div' only has a meaning when applied to a vector. Also the order of the operations is not interchangeable, thus

$$\operatorname{div}\mathbf{grad}\,V = \frac{\partial^2 V}{\partial x^2} + \frac{\partial^2 V}{\partial y^2} + \frac{\partial^2 V}{\partial z^2} = \nabla^2 V,$$

but $\mathbf{grad}\operatorname{div} V$ is meaningless.

Again, $\operatorname{div}\mathbf{grad}\,\mathbf{A}$ is meaningless, but

$$\mathbf{grad}\operatorname{div}\mathbf{A} = \left(\mathbf{i}\,\frac{\partial}{\partial x} + \mathbf{j}\,\frac{\partial}{\partial y} + \mathbf{k}\,\frac{\partial}{\partial z}\right)\left(\frac{\partial A_x}{\partial x} + \frac{\partial A_y}{\partial y} + \frac{\partial A_z}{\partial z}\right).$$

1·66. If we assume that the divergence as defined in **1·65** is independent of the shape of δv, it is easy to obtain the result (1) by taking for δv a rectangular parallelepiped with edges parallel to the axes and of lengths δx, δy, δz with P at its centre. The method is given in full in the author's *Electricity and Magnetism*, pp. 6, 7. We shall not reproduce it here, as in the next article we shall apply the same method to a more general case.

1·67. Curvilinear co-ordinates. Gradient and Divergence. Let the equations

$$f_1(x, y, z) = \alpha, \quad f_2(x, y, z) = \beta, \quad f_3(x, y, z) = \gamma \quad \ldots\ldots(1),$$

in which α, β, γ denote variable parameters, represent three families of orthogonal surfaces; i.e. surfaces which cut at right angles wherever they intersect one another.

We assume that one and only one surface of each family passes through any point (x, y, z), so that the parameters α, β, γ may be regarded as curvilinear co-ordinates of the point whose cartesian co-ordinates are x, y, z.

Let ds_1, ds_2, ds_3 denote elements of arc of the curves of intersection of the pairs of surfaces β, γ; γ, α and α, β. Then since α is the only parameter that changes along the curve

β, γ, it follows that ds_1 must be a multiple of $d\alpha$. We may therefore suppose that

$$ds_1 = h_1 d\alpha, \quad ds_2 = h_2 d\beta, \quad ds_3 = h_3 d\gamma \quad \ldots\ldots(2),$$

where h_1, h_2, h_3 are in general functions of the co-ordinates.*

Since the surfaces α, β, γ through any point are at right angles, a vector **A** at the point may be resolved into three components normal to these surfaces and denoted by A_α, A_β, A_γ.

Consider an element of volume dv bounded by surfaces $\alpha \pm \frac{1}{2}d\alpha$, $\beta \pm \frac{1}{2}d\beta$, $\gamma \pm \frac{1}{2}d\gamma$. Let the surface α cut this element of volume in the curvilinear rectangle $PQRS$ whose centre O is the point (α, β, γ).

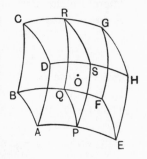

To find an expression for div **A** in curvilinear co-ordinates we have to find the flux of **A** out of dv. The component of **A** at O normal to $PQRS$ is A_α. Hence, if we assume what can easily be proved†, that the average value of a function over a small plane area is the value at the centroid, then the flux across $PQRS$ is A_α (area $PQRS$) $= A_\alpha . h_2 d\beta . h_3 d\gamma$.

* If the values of h_1, h_2, h_3 are required, they may be found by differentiating each of equations (1) with regard to α, β and γ, and solving the resulting equations for $\partial x/\partial \alpha$, $\partial x/\partial \beta$, etc. Then

$$ds_1{}^2 = \left\{ \left(\frac{\partial x}{\partial \alpha}\right)^2 + \left(\frac{\partial y}{\partial \alpha}\right)^2 + \left(\frac{\partial z}{\partial \alpha}\right)^2 \right\} d\alpha^2,$$

so that

$$h_1{}^2 = \left(\frac{\partial x}{\partial \alpha}\right)^2 + \left(\frac{\partial y}{\partial \alpha}\right)^2 + \left(\frac{\partial z}{\partial \alpha}\right)^2, \quad \text{etc.}$$

† If $f(x, y)$ be the function, S the area and $(\bar{x}, \bar{y})$ its centroid, we have

$$\int f(x, y) \, dS = \int f(\bar{x} + x', \bar{y} + y') \, dS$$

$$= \int \left\{ f(\bar{x}, \bar{y}) + x'\frac{\partial f}{\partial \bar{x}} + y'\frac{\partial f}{\partial \bar{y}} + \text{higher powers of } x', y' \right\} dS$$

$$= S\{f(\bar{x}, \bar{y}) + \epsilon^2\},$$

since

$$\int x' \, dS = \int y' \, dS = 0,$$

where ϵ is of order of the greatest linear dimension of S.

Now the parallel surfaces $ABCD$, $EFGH$ are the surfaces $\alpha - \tfrac{1}{2}d\alpha$ and $\alpha + \tfrac{1}{2}d\alpha$, and the outward flux of $\mathbf{A}$ across $ABCD$ is

$$-\left\{ A_\alpha h_2 h_3 \, d\beta d\gamma - \frac{1}{2}\frac{\partial}{\partial \alpha}(A_\alpha h_2 h_3 \, d\beta d\gamma)\, d\alpha + \epsilon_1 d\alpha d\beta d\gamma \right\},$$

where ϵ_1 is an infinitesimal which $\to 0$ as $d\alpha \to 0$.

Similarly the outward flux across the face $EFGH$ is

$$A_\alpha h_2 h_3 \, d\beta d\gamma + \frac{1}{2}\frac{\partial}{\partial \alpha}(A_\alpha h_2 h_3 \, d\beta d\gamma)\, d\alpha + \epsilon_2 d\alpha d\beta d\gamma,$$

where ϵ_2 has a like meaning to ϵ_1.

Therefore the contribution of this pair of faces to the flux of $\mathbf{A}$ out of dv is

$$\left\{ \frac{\partial}{\partial \alpha}(h_2 h_3 A_\alpha) + \epsilon \right\} d\alpha d\beta d\gamma,$$

where $\epsilon \to 0$ as $d\alpha \to 0$.

In like manner for the other pairs of faces of dv, so that the total outward flux of $\mathbf{A}$ is

$$\left\{ \frac{\partial}{\partial \alpha}(h_2 h_3 A_\alpha) + \frac{\partial}{\partial \beta}(h_3 h_1 A_\beta) + \frac{\partial}{\partial \gamma}(h_1 h_2 A_\gamma) + \eta \right\} d\alpha d\beta d\gamma,$$

where $\eta \to 0$ as $dv \to 0$.

Hence, since $\operatorname{div}\mathbf{A} = \lim\limits_{dv \to 0} \dfrac{\text{flux } \mathbf{A}}{dv}$, and $dv = h_1 h_2 h_3 \, d\alpha d\beta d\gamma$, therefore

$$\operatorname{div}\mathbf{A} = \frac{1}{h_1 h_2 h_3}\left\{ \frac{\partial}{\partial \alpha}(h_2 h_3 A_\alpha) + \frac{\partial}{\partial \beta}(h_3 h_1 A_\beta) + \frac{\partial}{\partial \gamma}(h_1 h_2 A_\gamma) \right\}$$
$$\dots\dots(3).$$

Further, when $\mathbf{A}$ is the gradient of a scalar function V, we have

$$A_\alpha,\ A_\beta,\ A_\gamma = \frac{\partial V}{\partial s_1},\quad \frac{\partial V}{\partial s_2},\quad \frac{\partial V}{\partial s_3} = \frac{1}{h_1}\frac{\partial V}{\partial \alpha},\quad \frac{1}{h_2}\frac{\partial V}{\partial \beta},\quad \frac{1}{h_3}\frac{\partial V}{\partial \gamma}$$
$$\dots\dots(4),$$

and by substituting (4) in (3), we have $\operatorname{div}\mathbf{grad}\, V$ or

$$\nabla^2 V = \frac{1}{h_1 h_2 h_3}\left\{ \frac{\partial}{\partial \alpha}\left(\frac{h_2 h_3}{h_1}\frac{\partial V}{\partial \alpha}\right) + \frac{\partial}{\partial \beta}\left(\frac{h_3 h_1}{h_2}\frac{\partial V}{\partial \beta}\right) + \frac{\partial}{\partial \gamma}\left(\frac{h_1 h_2}{h_3}\frac{\partial V}{\partial \gamma}\right) \right\}$$
$$\dots\dots(5).$$

1·68. Polar and cylindrical co-ordinates. When using polar co-ordinates r, θ, ϕ, the families of orthogonal surfaces are concentric spheres $r = \alpha$, cones $\theta = \beta$, and planes $\phi = \gamma$, and it is easy to see that

$$ds_1, \, ds_2, \, ds_3 \equiv dr, \, r\,d\theta, \, r\sin\theta\,d\phi,$$

so that $\qquad h_1, \, h_2, \, h_3 \equiv 1, \, r, \, r\sin\theta,$

and

$$\nabla^2 V = \frac{1}{r^2 \sin\theta}\left\{ \frac{\partial}{\partial r}\left(r^2 \sin\theta \frac{\partial V}{\partial r} \right) + \frac{\partial}{\partial \theta}\left(\sin\theta \frac{\partial V}{\partial \theta} \right) + \frac{\partial}{\partial \phi}\left(\frac{1}{\sin\theta}\frac{\partial V}{\partial \phi} \right) \right\},$$

or $\quad \nabla^2 V = \frac{1}{r^2}\frac{\partial}{\partial r}\left(r^2 \frac{\partial V}{\partial r} \right) + \frac{1}{r^2 \sin\theta}\frac{\partial}{\partial \theta}\left(\sin\theta \frac{\partial V}{\partial \theta} \right) + \frac{1}{r^2 \sin^2\theta}\frac{\partial^2 V}{\partial \phi^2}$

$$\ldots\ldots(1).$$

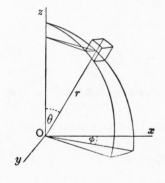 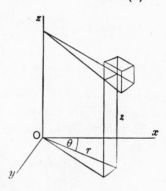

When using cylindrical co-ordinates r, θ, z, the families of orthogonal surfaces are co-axial cylinders $r = \alpha$, planes $\theta = \beta$, and planes $z = \gamma$, and in this case we have

$$ds_1, \, ds_2, \, ds_3 \equiv dr, \, r\,d\theta, \, dz,$$

so that $\qquad h_1, \, h_2, \, h_3 \equiv 1, \, r, \, 1,$

and $\qquad \nabla^2 V = \frac{1}{r}\left\{ \frac{\partial}{\partial r}\left(r\frac{\partial V}{\partial r} \right) + \frac{\partial}{\partial \theta}\left(\frac{1}{r}\frac{\partial V}{\partial \theta} \right) + \frac{\partial}{\partial z}\left(r\frac{\partial V}{\partial z} \right) \right\},$

or $\qquad \nabla^2 V = \frac{1}{r}\frac{\partial}{\partial r}\left(r\frac{\partial V}{\partial r} \right) + \frac{1}{r^2}\frac{\partial^2 V}{\partial \theta^2} + \frac{\partial^2 V}{\partial z^2} \ \ldots\ldots\ldots\ldots(2).$

Either of the results (1) and (2) can of course be obtained directly by applying the method of **1·67** to the appropriately shaped element of volume, and we suggest as an exercise to the student the independent verification of results (1) and (2) in this way.*

1·7. *If a vector field* **A** *is continuous throughout a region R bounded by a surface S, then the outward flux of* **A** *through the surface S is equal to the volume integral of div* **A** *throughout R.*

For we may subdivide R into a large number n of elements of volume δv_1, δv_2, ... δv_n. The total flux of **A** out of all these elements will be the outward flux across S; because over all boundaries which are not parts of S the flux will be taken twice in opposite directions.

Hence if S_r denotes the boundary surface of δv_r, we have

$$\int A_n dS = \sum_{r=1}^{n} \int A_n dS_r$$

$$= \sum_{r=1}^{n} (\text{div } \mathbf{A}\, \delta v_r + \epsilon_r \delta v_r),$$

where $\epsilon_r \to 0$ when $\delta v_r \to 0$. The latter step follows from the definition of div **A** as a limit.

Therefore, by increasing n so that the δv's tend to zero, we get

$$\int A_n dS \quad \text{or} \quad \int (\mathbf{A}\, d\mathbf{S}) = \int \text{div } \mathbf{A}\, dv \quad \ldots\ldots\ldots(1).$$

1·71. The theorem of **1·7** may be extended to apply to the case in which the region R contains a surface S' across which the vector field is discontinuous, i.e. there is a sudden change in **A** in crossing the surface S'.

Suppose that S' divides R into two parts R_1 and R_2 in each of which we assume that **A** is continuous and let A_{1n}, A_{2n} denote the normal components of **A** in R_1, R_2 close to S', taken positively towards S' in both regions. Then we can apply the theorem of **1·7** to each of the regions R_1, R_2 separately and add the results. This gives

$$\int_R \text{div } \mathbf{A}\ dv = \int_S A_n dS + \int_{S'} (A_{1n} + A_{2n})\, dS' \quad \ldots\ldots\ldots(1).$$

Across any surface at which **A** is continuous, we have, of course, $A_{2n} = -A_{1n}$ and the result reduces to **1·7** (1).

* See the author's *Electricity and Magnetism*, pp. 8, 9.

1·72. Green's Second Theorem. The theorem of **1·31** may be obtained in vector form from **1·7**, thus:

Let $\mathbf{A} = V \operatorname{grad} V'$, where V, V' are scalar functions which with their first and second derivatives are finite and continuous through the region bounded by the surface S.

Then
$$\int (\mathbf{A}\, d\mathbf{S}) = \int V \,(\operatorname{grad} V' d\mathbf{S}),$$

and
$$\operatorname{div} \mathbf{A} = \nabla \mathbf{A} = \nabla\, (V \nabla V')$$
$$= \nabla V \nabla V' + V \nabla^2 V'$$
$$= \operatorname{grad} V \operatorname{grad} V' + V \operatorname{div} \operatorname{grad} V'.$$

Therefore **1·7** (1) is equivalent to

$$\int \operatorname{grad} V \operatorname{grad} V'\, dv = \int V\, (\operatorname{grad} V' d\mathbf{S}) - \int V \operatorname{div} \operatorname{grad} V' dv$$
$$\dots\dots(1),$$

or if we use the Hamiltonian operator

$$\int \nabla V \nabla V' dv = \int V\, (\nabla V' d\mathbf{S}) - \int V \nabla^2 V' dv \quad \dots\dots(2);$$

and the symmetry of the left-hand side shews that V and V' might be interchanged on the right-hand side. The reader will have no difficulty in identifying (1) and (2) with **1·31** (2).

1·8. Convergence of volume integrals. In the theory of attraction and kindred subjects we frequently have to consider integrals of the form $\int \dfrac{\rho\, dv}{r^n}$, where dv denotes an element of volume and ρ a scalar function of position, and the region of integration R is bounded by a surface S which includes the origin O from which the radius vector r to a point of dv is measured. It follows that the integrand becomes infinite within the range of integration, and we have to investigate the circumstances under which the integral can be said to have a meaning.

Draw a closed surface S_1 inside S and surrounding O, then so far as the subject of integration depends on r it is finite between S_1 and S, and $\int_{S_1}^{S} \dfrac{\rho\, dv}{r^n}$ has a definite value, where the limits indicate that the integral is taken through the volume bounded by S and S_1. If as S_1 contracts upon the point O the last integral tends to a definite limit which is independent of the shape of S_1, then

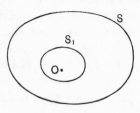

$$\lim_{S_1 \to 0} \int_{S_1}^{S} \frac{\rho\, dv}{r^n}$$

is defined to be the value of the given integral through R and the integral is said to *converge*. If the integral does not converge, it is said to *diverge*.

A condition for convergence is that corresponding to any small positive number ϵ there must be a surface S_1, defined as above, such that for all surfaces S_2 inside S_1 and surrounding O

$$\left| \int_{S_2}^{S_1} \frac{\rho\, dv}{r^n} \right| < \epsilon \quad \dots\dots\dots\dots\dots\dots(1).$$

For if a limit A exists we can choose S_1 such that

$$\left| \int_{S_1}^{S} \frac{\rho\, dv}{r^n} - A \right| < \tfrac{1}{2}\epsilon, \quad \text{and} \quad \left| \int_{S_2}^{S} \frac{\rho\, dv}{r^n} - A \right| < \tfrac{1}{2}\epsilon,$$

and therefore $\quad \left| \int_{S_2}^{S_1} \frac{\rho\, dv}{r^n} \right| = \left| \int_{S_2}^{S} - \int_{S_1}^{S} \frac{\rho\, dv}{r^n} \right| < \epsilon.$

Conversely, if condition (1) holds, then a limit exists, for by taking S_1 small enough we can make the fluctuation of the value of the integral for different cavities inside S_1 as small as we please and therefore small without limit.

Now let M denote the greatest value of ρ within a finite region round O. Let η denote the greatest value of r on S_1 and r_2 the least value of r on S_2. Let Σ be the sphere of radius η and σ the sphere of radius r_2. Then

$$\left| \int_{S_2}^{S_1} \frac{\rho}{r^n}\, dv \right| \leqslant M \left| \int_{\sigma}^{\Sigma} \frac{dv}{r^n} \right| \quad \dots\dots\dots\dots(2);$$

but, as in **1·42**, we may put $dv = r^2 d\omega\, dr$, so that the contribution to the last integral of the cone of solid angle $d\omega$, when $n < 3$, is

$$d\omega \int_{r_2}^{\eta} r^{2-n} dr \quad \text{or} \quad \frac{d\omega}{3-n} (\eta^{3-n} - r_2^{3-n}) < \frac{\eta^{3-n} d\omega}{3-n}.$$

Consequently $\int_{\sigma}^{\Sigma} \dfrac{dv}{r^n} < \dfrac{\eta^{3-n}}{3-n} \int d\omega \quad \text{or} \quad < \dfrac{4\pi\eta^{3-n}}{3-n}$,

and $\qquad \left| \displaystyle\int_{S_2}^{S_1} \dfrac{\rho\, dv}{r^n} \right| < \dfrac{4\pi M \eta^{3-n}}{3-n}.$

Hence, provided $n < 3$, for any arbitrarily chosen positive number ϵ however small we have only to take η less than $\{(3-n)\,\epsilon/4\pi M\}^{\frac{1}{3-n}}$ in order to get a surface S_1 such that for all surfaces S_2 inside S_1

$$\left| \int_{S_2}^{S_1} \frac{\rho\, dv}{r^n} \right| < \epsilon.$$

Consequently the given integral is convergent if $n < 3$.

We note that the form of the outer boundary S does not come into consideration, because it is only in the vicinity of the origin O that the integrand tends to become infinite.

It can be shewn that for $n \geqslant 3$ the integral is divergent.[*]

[*] For further information on the subject see J. G. Leathem's Tract on *Volume and Surface Integrals used in Physics*, Camb. Univ. Press, 1905.

Chapter II

GRAVITATIONAL ATTRACTION AND POTENTIAL. SIMPLE APPLICATIONS

2·1. The Law of Gravitation. The law discovered by Newton* is as follows: *Every particle in the universe attracts every other particle with a force which is directly proportional to the product of the masses of the particles and inversely proportional to the square of the distance between them.*†

Thus if m, m' denote the masses of two particles and r their distance apart, the force of attraction between them is $\gamma \dfrac{mm'}{r^2}$; where γ is known as the **gravitation constant**, and measures the attraction of two particles of unit mass at unit distance apart.

It is evident that some limitation must be imposed upon the dimensions of the particles in order to avoid a difficulty in defining the distance between them. For this purpose we may define **a material particle** as a body so small that, for the purposes of our investigation, the distance between its different parts may be neglected.

We shall prove later that a uniform solid sphere attracts external particles as though its mass were concentrated at its centre; and this may be regarded as some justification for

---

* Sir Isaac Newton (1642–1727), Lucasian Professor of Mathematics at Cambridge and Fellow of Trinity College. His great work *Philosophiae Naturalis Principia Mathematica* was published in 1687 by the Royal Society, then under the Presidency of Samuel Pepys. He made discoveries in Optics and invented his *Methodus Fluxionum* which constituted the foundation of the Calculus. In later life he left Cambridge for London, becoming President of the Royal Society and Master of the Mint.

† An outline of the process of reasoning by deduction, from Kepler's laws of planetary motion through the propositions of Newton's *Principia* to the law of gravitation, is given in the author's *Dynamics*, Part I, pp. 167–8.

the hypothesis we propose to make, namely, that for the purpose of calculating attractions we may regard particles as quantities of matter concentrated at points.

2·11. Numerical value of γ.

An approximate value for γ may be determined thus: Regarding the earth as a sphere of radius r and assuming it to attract a small body on its surface as though its mass were collected at its centre, then if E denotes the mass of the earth in grammes and M that of the body,

$$\gamma \frac{EM}{r^2} = \text{weight of the body} = Mg \qquad \ldots\ldots(1).$$

So that, if ρ denotes the mean density of the earth, $E = \frac{4}{3}\pi\rho r^3$, and from (1),

$$\gamma = \frac{3g}{4\pi\rho r} \qquad \ldots\ldots\ldots\ldots\ldots\ldots\ldots(2).$$

But, in c.g.s. units,

$$g = 981 \text{ cm. sec.}^{-2}, \quad \rho = 5\cdot67 \text{ approx.}$$

and $\quad \frac{1}{2}\pi r = $ a quadrant of the earth's circumference

$$= 10^9 \text{ cm.}$$

Therefore $\quad \gamma = \dfrac{3 \times 981}{8 \times 10^9 \times 5\cdot67} = \dfrac{1}{15,500,000}$ approx. $\ldots\ldots(3)$,

when c.g.s. units are used.

The mean density of the earth has been determined by a series of experiments in which the attraction of the earth on a given body has been compared with the attraction on the same body of another body of known mass.*

2·12. Astronomical units.

We can avoid the continuous repetition of the constant γ throughout our work by choosing units such that $\gamma = 1$. Such units are called **astronomical** or **theoretical units**.

We observe that the acceleration f produced by the attraction of a particle of mass m on a particle at a distance r is given by $f = \gamma \dfrac{m}{r^2}$; so that $\gamma = 1$ when f, m and r are all unity.

---

* See *Encycl. Brit.* article 'Gravitation'.

Hence the astronomical unit of mass is the mass of a particle which by its attraction produces unit acceleration at unit distance.

We can find the astronomical unit of mass in grammes by taking the above formula for acceleration, which holds good in all systems of units, and putting $r = 1$ cm. and $f = 1$ cm. sec.$^{-2}$, so that $\gamma m = 1$, or $m = 1/\gamma = 15,500,000$ grammes.

In what follows we shall omit the constant γ, save that in the collections of examples from examination papers we shall print the questions in the form in which they were set, in order that readers may become familiar with the use of γ.

2·2. Field of force. When we speak of the attraction of a body or system of particles at a point external to itself, we mean the force of attraction which the body or system would exert on a particle of unit mass placed at the point. There must be a definite value (including zero at a point of equilibrium) for this force at every point at which a particle can be placed. Thus we arrive at the conception of a **field of force**, or region of space with every point of which there is associated a force which is definite in magnitude and direction.

Inasmuch as a force belongs to the class of physical magnitudes known as vectors, the field of attraction of any system of bodies is an example of a **vector field**.

2·21. Attraction of a system of particles. It is part of the Newtonian theory of attraction that the force exerted at a point by a system of particles is the vector sum of the forces exerted by each of the particles separately. This is called the principle of superposition of fields of force.

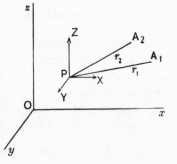

Let particles of masses m_1, $m_2, m_3, \ldots$ be situated at points A_1, A_2, A_3, $\ldots$ whose co-ordinates referred to rectangular axes are a_1, b_1, c_1; a_2, b_2, c_2;

a_3, b_3, c_3; .... It is required to find expressions for the components X, Y, Z of the attraction of the system of particles at any point P whose co-ordinates are x, y, z.

Let r_1, r_2, r_3, ... denote the distance PA_1, PA_2, PA_3, ...

The attraction at P of m_1 at A_1 is m_1/r_1^2 directed along PA_1; and PA_1 has direction cosines $\dfrac{a_1-x}{r_1}, \dfrac{b_1-y}{r_1}, \dfrac{c_1-z}{r_1}$, where

$$r_1^2 = (a_1-x)^2 + (b_1-y)^2 + (c_1-z)^2.$$

Therefore the particle m_1 exerts at P a force whose components parallel to the axes are

$$\frac{m_1(a_1-x)}{r_1^3}, \quad \frac{m_1(b_1-y)}{r_1^3}, \quad \frac{m_1(c_1-z)}{r_1^3}.$$

The other particles m_2, m_3, ... make like contributions, so that, by the principle of superposition of fields of force, we have

$$X = \Sigma\left\{\frac{m_s(a_s-x)}{r_s^3}\right\}, \quad Y = \Sigma\left\{\frac{m_s(b_s-y)}{r_s^3}\right\}, \quad Z = \Sigma\left\{\frac{m_s(c_s-z)}{r_s^3}\right\}$$
$$\ldots\ldots(1),$$

where $s = 1$, 2, 3, ... and the summation extends to all the attracting particles.

2·22. Potential. Let us now define a function V by the formula

$$V = \Sigma\frac{m_s}{r_s} \qquad\qquad\ldots\ldots\ldots\ldots\ldots\ldots(1),$$

where $\qquad r_s^2 = (a_s-x)^2 + (b_s-y)^2 + (c_s-z)^2 \quad \ldots\ldots\ldots(2),$

in the notation of 2·21.

Thus defined, V is a function related to a system of attracting particles having a definite value at every point P of space external to the particles. It is a function of the co-ordinates (x, y, z) of P and is clearly a single-valued function, in the sense that it cannot have more than one value at each point P; for it represents simply the sum of the masses of the separate particles divided by their respective distances from P. Further, though we have expressed V as a function of x, y, z, it clearly represents a sum which does not depend on the particular system of axes of reference.

Now, by differentiation, we have

$$\frac{\partial V}{\partial x} = -\Sigma \frac{m_s}{r_s^2} \frac{\partial r_s}{\partial x}.$$

But, from (2), $r_s \frac{\partial r_s}{\partial x} = -(a_s - x),$

so that $\frac{\partial V}{\partial x} = \Sigma \left\{ \frac{m_s (a_s - x)}{r_s^3} \right\},$

and by 2·21 (1) this is equal to X. Similarly

$$\frac{\partial V}{\partial y} = Y \quad \text{and} \quad \frac{\partial V}{\partial z} = Z. \quad \ldots\ldots\ldots\ldots(3).$$

The function V defined by (1) is called the **potential of the attracting particles**, or the **potential of the field of force**. We have proved that the derivatives of V with regard to x, y, z give the components of attraction at P in the directions of the axes. And since the directions of the axes can be chosen arbitrarily, it follows that the space-derivative of V in any direction gives the component of attraction in that direction. This may be verified thus: let $\partial/\partial s$ denote a differentiation in a direction ds whose direction cosines are l, m, n or $\frac{\partial x}{\partial s}, \frac{\partial y}{\partial s}, \frac{\partial z}{\partial s}$. Then

$$\frac{\partial V}{\partial s} = \frac{\partial V}{\partial x} \frac{\partial x}{\partial s} + \frac{\partial V}{\partial y} \frac{\partial y}{\partial s} + \frac{\partial V}{\partial z} \frac{\partial z}{\partial s}$$

$$= l \frac{\partial V}{\partial x} + m \frac{\partial V}{\partial y} + n \frac{\partial V}{\partial z}$$

$$= lX + mY + nZ$$

$$= \text{force component in direction } ds \quad \ldots\ldots(4).$$

In the language of vectors the **force of attraction is the gradient of the potential** (1·5), or, if **R** denotes the resultant attraction at a point at which the potential is V,

$$\mathbf{R} = \text{grad } V \quad \ldots\ldots\ldots\ldots\ldots(5).$$

The potential function V plays an important part in the theory of attractions, since, as we have just seen, if the potential of any given distribution of matter can be determined the attraction at any point can be found immediately by differentiating the potential.

2·23. Physical interpretation of potential. With the notation of **2·22** we have, for the total differential of the potential,

$$dV = \frac{\partial V}{\partial x}\,dx + \frac{\partial V}{\partial y}\,dy + \frac{\partial V}{\partial z}\,dz$$

$$= X\,dx + Y\,dy + Z\,dz,$$

from **2·22** (3).

Hence, by integrating along any path from P to Q, we get

$$V_Q - V_P = \int_P^Q \left(X\frac{\partial x}{\partial s} + Y\frac{\partial y}{\partial s} + Z\frac{\partial z}{\partial s} \right) ds \quad \ldots\ldots\ldots(1).$$

But the integral represents the work which the forces of attraction would perform upon a particle of unit mass as it moved along this path from P to Q. This gives us a measure of potential in terms of *work per unit mass*, in that *the potential at any point Q exceeds the potential at any other point P by the work which the forces of attraction would perform upon a particle of unit mass as it moves along any path from P to Q.*

It is clear that the addition of a constant to the potential will not affect the values of the force components since these are obtained by differentiating the potential. Also, when the potential is determined by integration, from known force components, as in (1) above, the constant of integration may be so chosen as to make the potential vanish at an infinite distance from the attracting matter, since this will accord with the definition of **2·22** (1). On this hypothesis we see that the *potential at a given point due to a given attracting system is the work that would be done by the attractions of the system on a particle of unit mass as it moves along any path from an infinite distance up to the point considered.*

We have seen that the definition of potential as $\Sigma\,(m/r)$ leads to this expression for potential in terms of work done per unit mass, and it is easy to demonstrate the converse, thus:

Let m be the mass of a typical particle of the system situated at the point A. Let $PP' = ds$ be an element of any path from an infinite distance to the point Q, and let $AP = r$, and $AP' = r + dr$. Then so far as the field of force depends upon the particle m at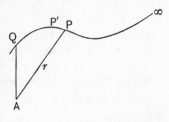

A its value at P is m/r^2 directed along PA; and the work done by this force on the unit particle as it moves from P to P' is $\dfrac{m}{r^2}\left(\dfrac{-dr}{ds}\right)ds$, or $\dfrac{-m\,dr}{r^2}$. Hence the total work done by the attraction of the particle m on a unit particle moving from an infinite distance to the point Q is

$$-\int_{\infty}^{AQ}\frac{m}{r^2}\,dr \quad \text{or} \quad \frac{m}{AQ}.$$

By the principle of superposition of fields, the total work done by the attractions of all the particles of the system is obtained by adding for their separate effects, so that

$$V = \Sigma\,\frac{m}{AQ}$$

gives the potential at Q; where the formula represents the sum of the masses of the separate particles each divided by its distance from Q. The interchangeability of the two definitions of potential is thus completely established.

2·24. Dimensions. *Gravitational potential* must not be confused with *potential energy*. They have different physical dimensions; because the dimensions of *potential energy* are those of *work*, i.e. $\mathbf{ML^2T^{-2}}$, in terms of the fundamental units of mass, space and time; but the dimensions of *gravitational potential* as explained in **2·23** are those of *work per unit mass*, i.e. $\mathbf{L^2T^{-2}}$, or the square of a velocity.

In this connection it is important to remember that we are using astronomical units (2·12) and omitting the gravitational constant γ, and though this does not affect the argument when potential is defined as work per unit mass, yet when we use the formula for potential of 2·22 (1) if we want to find the dimensions of V we must restore the constant γ and write $V = \gamma \Sigma\,(m/r)$, because γ has dimensions. By definition $\gamma \dfrac{mm'}{r^2}$ represents a force and is therefore of dimensions $\mathbf{MLT^{-2}}$, so that the dimensions of γ are $\mathbf{M^{-1}L^3T^{-2}}$, and the dimensions of $\gamma \dfrac{m}{r}$ are $\mathbf{L^2T^{-2}}$ as above.

It is important also to note that though potential energy decreases when work is done, gravitational potential increases.

2·25. The definition of potential in terms of work (2·23) leads at once to the relation
$$\mathbf{R} = \operatorname{grad} V$$
of 2·22. Thus if P, P' are any two neighbouring points at a distance ds apart, V, $V + dV$ denote the potentials at P, P' and F the component of attraction at P in direction PP', we have

$(V + dV) - V = $ increment in potential from P to P'

$\qquad\qquad$ = work done *by* attractions on unit mass passing from P to P'

$\qquad\qquad$ = $F ds$.

Therefore $F = \dfrac{\partial V}{\partial s}$, and this being true for components of attraction in every direction round P it follows that
$$\mathbf{R} = \operatorname{grad} V.$$

2·26. Equipotential surfaces and Lines of force. Regarding the potential V of a given attracting system as a function of co-ordinates x, y, z, the equation
$$V\,(x, y, z) = \text{const.}$$
represents a surface over which the potential is constant. Such surfaces are called **equipotential surfaces**. It is evident from the definition of potential that only one such surface passes through any point of space, so that no two equipotential surfaces can intersect. Also since V has a constant value over such a surface no work would be done by

the attractions on a particle moving on such a surface, therefore at every point the resultant attraction is normal to the equipotential surface through the point. This is also obvious from the relation $\mathbf{R} = \operatorname{grad} V$ (1·5).

A line so drawn that its tangent at each point is in the direction of the resultant attraction at the point is called a **line of force**. It is evident that a line of force is at right angles to the equipotential surfaces at all their points of intersection; and, conversely, a surface which cuts all lines of force at right angles must be an equipotential surface, because at no point on the surface is there a component of force tangential to the surface, so that no work would be done on a particle moving on the surface and there could therefore be no variation in the potential.

2·3. Continuous bodies. We have now to pass from the attraction components and potential of a system of separate particles to the attraction components and potential of distributions of matter regarded as *continuous bodies*. We remark in the first place that such bodies are not really continuous but consist of separate molecules and ultimately of indivisible electrons. By the principle of superposition therefore there should be no difficulty in obtaining the attraction components and potential of such a body provided that we have a means of summing the contributions of the separate particles. It is natural to look to integration to effect the summation, but when we consider what the process of integration involves, we find that it does not fit the physical conditions of the problem precisely, and it is only by giving a special interpretation to our conception of 'body' that we can justify the use of integration. Thus it is usual to represent the potential of a continuous body by a volume integral $\int \rho \, dv / r$, where dv denotes an element of volume of the body at a distance r from an external point P and ρ denotes the density of matter in dv. The process of integration implies that the density of the body is continuous; but this is not the case for a real body with a molecular structure. To justify the use of such an integral it is necessary therefore to suppose that it is applied to a hypothetical *continuous* distribution of matter occupying the same region as the body and having at each point a suitably chosen density; this density being found by considering a small but finite volume surrounding the point and taking the average through this small volume of the masses of the particles of the real body contained therein.

2·4. Before proceeding further with the general theory of the subject we shall calculate the attraction components and the potential of some simple bodies.

Attraction of a uniform straight rod. Let m denote the mass per unit length of a uniform rod AB. It is required to find the components of attraction of AB at an external point P.

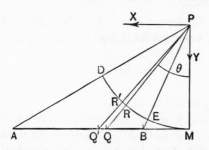

Let the perpendicular from P to AB meet AB in M, which for simplicity we take on AB produced though this is not necessary for the argument. Let $MP = p$. Consider an element QQ' of the rod, where $MQ = x$ and $QQ' = dx$, and let the angle MPQ be θ. Then $x = p \tan \theta$, and the mass of the element QQ' of the rod is $m\,dx = mp \sec^2 \theta\,d\theta$.

The attraction at P of this element is therefore

$$\frac{mp \sec^2 \theta\,d\theta}{PQ^2} \quad \text{along } PQ,$$

or
$$\frac{mp \sec^2 \theta\,d\theta}{p^2 \sec^2 \theta} = \frac{m\,d\theta}{p} \quad \text{along } PQ \quad \ldots\ldots\ldots\ldots(1).$$

Denoting the components of attraction of the rod parallel and perpendicular to BA by X and Y, and the angles MPA,

MPB by α, β, we have

$$X = \int_\beta^\alpha \frac{m}{p}\sin\theta\,d\theta = \frac{m}{p}(\cos\beta - \cos\alpha)$$

$$= \frac{2m}{p}\sin\tfrac{1}{2}(\alpha+\beta)\sin\tfrac{1}{2}(\alpha-\beta) \quad\ldots\ldots(2)$$

and $\quad Y = \int_\beta^\alpha \frac{m}{p}\cos\theta\,d\theta = \frac{m}{p}(\sin\alpha - \sin\beta)$

$$= \frac{2m}{p}\sin\tfrac{1}{2}(\alpha-\beta)\cos\tfrac{1}{2}(\alpha+\beta) \quad\ldots\ldots(3).$$

The resultant attraction is therefore $\dfrac{2m}{p}\sin\tfrac{1}{2}(\alpha-\beta)$, or $\dfrac{2m}{p}\sin\tfrac{1}{2}APB$, and it makes an angle $\tan^{-1}\dfrac{X}{Y}$ or $\tfrac{1}{2}(\alpha+\beta)$ with PM, i.e. it acts along the bisector of the angle APB.

Another convenient form for the component *parallel* to the rod is

$$X = \frac{m}{p}(\cos\beta - \cos\alpha) = \frac{m}{PB} - \frac{m}{PA} \quad\ldots\ldots\ldots\ldots(4)$$

in the sense parallel to BA.

We note that if a circle of centre P and radius PM cuts PA, PQ', PQ, PB in D, R', R, E, then the attraction at P of the element RR' of a rod in the form of a circular arc DE of the same line density (mass per unit length) as AB is

$$\frac{mp\,d\theta}{p^2} = \frac{m\,d\theta}{p} = \text{attraction of } QQ'.$$

Hence this circular arc DE exerts the same attraction as the rod AB.

Cor. If the rod is infinitely long, the angle APB is two right angles and the resultant attraction is $2m/p$ perpendicular to the rod. It would appear from this result that if the attracted particle were close to the rod the attraction would be infinite; but this conclusion is not justified because in the foregoing argument we assumed that every point of an element QQ' of the rod was at the same distance from the point P, and for this to be true when P is close to the rod it would be necessary for the rod to have no thickness. To find the attraction at a point close to a rod of finite thickness it will be necessary to take account of the cross-section of the rod. The attraction of a uniform long rod whose cross-section is a circle is found in **2·42** below.

2·41. Potential of a uniform straight rod. With the notation of **2·4** the potential at P is given by

$$V = \int \frac{m\,dx}{PQ}$$

$$= \int_{\beta}^{\alpha} \frac{mp\sec^2\theta\,d\theta}{p\sec\theta}$$

$$= m\left[\log\tan\tfrac{1}{2}(\theta+\tfrac{1}{2}\pi)\right]_{\beta}^{\alpha}$$

$$= m\log\cot\tfrac{1}{2}PAB\cot\tfrac{1}{2}PBA \qquad \ldots\ldots\ldots(1).$$

If $2l$ denotes the length of the rod AB and $PA = r$, $PB = r'$, we easily find from (1) that*

$$V = m\log\frac{r+r'+2l}{r+r'-2l} \qquad \ldots\ldots\ldots\ldots\ldots(2).$$

It follows that if A, B are foci of an ellipse passing through P and $2a$ is its major axis,

$$V = m\log\frac{a+l}{a-l} \qquad \ldots\ldots\ldots\ldots\ldots(3),$$

or $$V = m\log\frac{1+e}{1-e} \qquad \ldots\ldots\ldots\ldots\ldots(4),$$

where e denotes the eccentricity of the ellipse.

Hence the potential is constant over any prolate spheroid of which A, B are the foci; i.e. this family of confocal prolate spheroids are the equipotential surfaces. Since the normal to an ellipse at any point bisects the angle between the focal

---

* If $r+r'+2l = 2s$,

$$\cot\tfrac{1}{2}PAB\cot\tfrac{1}{2}PBA = \sqrt{\left\{\frac{s(s-r')}{(s-r)(s-2l)}\cdot\frac{s(s-r)}{(s-r')(s-2l)}\right\}}$$

$$= \frac{s}{s-2l} = \frac{r+r'+2l}{r+r'-2l}.$$

distances and the resultant attraction at a point is normal to the equipotential surface, it follows that the resultant attraction at P bisects the angle APB, as we found in 2·4.

If the rod be of great length and P in the neighbourhood of its centre, then in (2) we may put $r+r' = 2\sqrt{(l^2+p^2)}$, where p is small compared with l. This gives

$$V = m \log \frac{\sqrt{(l^2+p^2)}+l}{\sqrt{(l^2+p^2)}-l}$$

$$= 2m \log \frac{\sqrt{(l^2+p^2)}+l}{p}$$

$$= 2m \log \left\{ \left(2l + \frac{p^2}{2l} \right) \Big/ p \right\},$$

or neglecting $(p/l)^2$,

$$V = 2m \log 2l - 2m \log p \quad \ldots\ldots\ldots\ldots(5).$$

By differentiating the last result we get for the attraction of the rod in the direction p increasing, $\dfrac{\partial V}{\partial p} = -\dfrac{2m}{p}$, agreeing with 2·4 Cor.; and conversely, if we put $\dfrac{\partial V}{\partial p} = -\dfrac{2m}{p}$ and integrate, we find for the potential $V = C - 2m \log p$, in agreement with (5).

2·42. Attraction and potential of a long thin uniform circular cylinder.

Take a cross-section of the cylinder about the middle of its length. Let O be the centre of the cross-section and P any point inside it. Through P draw chords QPR, $Q'PR'$ making a small angle $d\theta$ with one another, and intercepting small arcs QQ', RR' on the circle. We can conceive the cylinder to be composed of long parallel rods, and take QQ', RR' as the cross-sections of two of them. Then if m 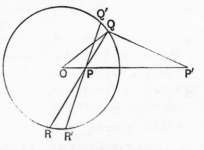 denotes the surface density (mass per unit area) of the cylinder, mQQ' and mRR' denote the mass per unit length of the two rods. Let the equal angles OQP, ORP be denoted by ϕ. Then using the result of

2·4 Cor. for the attraction of a long rod, we have

attraction at P due to rod through QQ'

$$= \frac{2mQQ'}{PQ} = 2m\sec\phi\, d\theta \quad \text{along } PQ,$$

attraction at P due to rod through RR'

$$= \frac{2mRR'}{PR} = 2m\sec\phi\, d\theta \quad \text{along } PR.$$

Hence these two rods exert equal and opposite attractions at P; and by dividing up the whole cylinder into similar pairs of rods we see that its resultant attraction at any internal point is zero.

It follows that the potential must be constant at all points inside the cylinder sufficiently far from its ends, and is therefore equal to the potential at O, and by **2·41 (5)** the potential due to the rod QQ' is

$$2mQQ'\log 2l - 2mQQ'\log OQ,$$

where $2l$ is the length of the cylinder. Therefore, for the whole potential,

$$V = 2M\log 2l - 2M\log a \quad \dots\dots\dots\dots\dots(1),$$

where $M = 2\pi ma$, a being the radius.

Again, for an external point, take P' the inverse of P, so that $OP.OP' = a^2$, and, by similar triangles, $OP'Q = OQP = \phi = OP'R$. Then the attraction at P' of the rod through QQ'

$$= \frac{2mQQ'}{P'Q} = \frac{2mPQ\, d\theta \sec\phi}{P'Q} = 2m\, d\theta \sec\phi \cdot \frac{a}{OP'} \quad \text{along } P'Q.$$

But the resultant attraction of the cylinder at P' is clearly along $P'O$, and by resolving the attraction of the rod through QQ' in this direction we get $\dfrac{2ma}{OP'}d\theta$. Similarly, other rods give like results and the whole attraction at P' is therefore $\dfrac{4\pi ma}{OP'} = \dfrac{2M}{OP'}$; and we observe that this is the same as if the whole mass of the cylinder were condensed into a rod of equal mass along the axis of the cylinder.

To find the potential, we may put r for OP' and write

$$\frac{dV}{dr} = -\frac{2M}{r},$$

so that $\qquad\qquad V = \text{const.} - 2M\log r,$

where, in order that V may take the form (1) when $r = a$, **(3·7)**, the constant is $2M\log 2l$, so that

$$V = 2M\log 2l - 2M\log r.$$

It is clear that the same results will be true for the attraction and potential of a uniform long solid cylinder *at external points*, for the cylinder can be divided up into coaxial thin cylinders for each of which the results are true.

2·5. Attraction of a uniform circular disc at a point on its axis. Let O be the centre of the disc, P a point on its axis Oz at a distance z from O, and let m denote the surface density or mass per unit area. Divide the disc into concentric rings. Let $OQ = x$ be the radius and $QQ' = dx$ the breadth of one of these rings. The mass of the ring is $2\pi m x\, dx$, and the attraction at P of each element is got by dividing the mass of the element by PQ^2. But the resultant attraction of the ring is clearly in the direction PO, so that its magnitude is

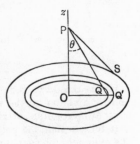

$$\frac{2\pi m x\, dx}{PQ^2} \cos\theta,$$

where θ is the angle OPQ. But $x = z\tan\theta$, so if α is the angle which a radius of the disc subtends at P, we have for the whole attraction of the disc

$$\int_0^\alpha \frac{2\pi m z^2 \tan\theta \sec^2\theta \cos\theta}{z^2 \sec^2\theta}\, d\theta = 2\pi m\,(1 - \cos\alpha) \;\ldots(1).$$

For an infinite plate we may put $\alpha = \tfrac{1}{2}\pi$, so that the attraction of an infinite plate is $2\pi m$ at right angles to itself. But we may not infer that the attraction of a circular disc at a point on its axis close to itself is $2\pi m$ unless the thickness of the plate is negligible, so that the particles of a ring in the immediate vicinity of the attracted particle may be regarded as equidistant from it.

A frustum of a right circular cone bounded by planes at right angles to its axis may be regarded as a pile of discs all subtending the same angle at the vertex, and the attraction at the vertex of the cone due to such a frustum is easily seen to be $2\pi m t\,(1 - \cos\alpha)$, where t denotes the thickness of the frustum, m the mass of unit volume, and α the semi-angle of the cone.

2·51. Potential at any point on the axis of a uniform circular disc. With the notation of 2·5, we have

$$V = \int_0^a \frac{2\pi m x\, dx}{PQ}, \quad \text{where } a \text{ is the radius of the disc,}$$

$$= 2\pi m \int_0^a \frac{x\, dx}{\sqrt{(z^2 + x^2)}} = 2\pi m \{\sqrt{(z^2 + a^2)} - z\} \ldots\ldots\ldots(1).$$

We note that this gives for the attraction in direction PO

$$-\frac{dV}{dz} = 2\pi m \left\{ 1 - \frac{z}{\sqrt{(z^2 + a^2)}} \right\}$$

in agreement with **2·5** (1). But if we find the potential from the attraction by integrating the last formula, we get

$$V = 2\pi m \{\sqrt{(z^2 + a^2)} - z\} + C,$$

where C is a constant of integration. In the physical definition of potential in terms of work, it is assumed that the potential is zero at an infinite distance from the attracting matter. And if in the case under consideration V is to vanish when $z \to \infty$, we must have $C = 0$.

If z/a is small, then by (1)

$$V = C - 2\pi m z, \quad \ldots\ldots\ldots\ldots\ldots\ldots(2),$$

where $C = 2\pi m a$, a constant.

It should be noted that the formula (1) is equivalent to $V = 2\pi m\,(SP - OP)$ [fig. **2·5**], and that this will give the value of V on either side of the disc if SP, OP denote numerical lengths. And for negative values of z, instead of (1) we should have

$$V = 2\pi m \{\sqrt{(z^2 + a^2)} + z\} \quad \ldots\ldots\ldots\ldots(3).$$

The expressions (1) and (3) are both comprised in the formula

$$V = 2\pi m \{\sqrt{(z^2 + a^2)} - |z|\}.$$

2·52. Attraction at a point on the axis of a uniform circular cylinder. Let l be the length, a the radius and ρ the density of the cylinder. Let the point O at which the attraction is to be found be at a distance c from the nearer end. The cylinder may be regarded as a pile of discs, and a disc of thickness dx at a distance x from O has a mass $\rho\, dx$ per unit area, so that its attraction is

$$2\pi\rho\left\{1-\frac{x}{\sqrt{(x^2+a^2)}}\right\}dx,$$

and the whole attraction

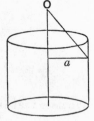

$$=2\pi\rho\int_c^{l+c}\left\{1-\frac{x}{\sqrt{(x^2+a^2)}}\right\}dx$$

$$=2\pi\rho\{l-\sqrt{[(l+c)^2+a^2]}+\sqrt{(c^2+a^2)}\}.$$

If the cylinder is of infinite length in one direction we have to find the limit of the last expression as $l\to\infty$, i.e. the limit of

$$2\pi\rho\left\{l-l\left(1+\frac{2c}{l}+\frac{c^2+a^2}{l^2}\right)^{\frac12}+\sqrt{(c^2+a^2)}\right\}\quad\text{as }l\to\infty$$

$$=2\pi\rho\{\sqrt{(c^2+a^2)}-c\}.$$

If the particle be at the centre of the end of such a cylinder the attraction is $2\pi\rho a$.

2·6. Use of solid angles.

There are many applications of the theory of attraction in which calculations are simplified by the use of the solid angle. For example, *to find the component of attraction perpendicular to itself produced by a plane plate of any form.*

Let dS be an element of area of the plate at a distance r from the point O and subtending a solid angle $d\omega$ at O. Then if m denotes the mass per unit area of the plate, the attraction of this element at O is $m\, dS/r^2$, and resolving this at right angles to the plate we get $m\, dS\cos\theta/r^2$, where θ is the inclination of r to the normal. But $dS\cos\theta/r^2=d\omega$ (**1·4**). Therefore $m\, d\omega$ is the contribution of this element dS of the plate to its whole

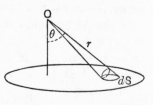

attraction at O at right angles to itself; and the attraction of the whole plate in the same direction is $m\omega$, where ω is the solid angle which the plate subtends at O.

2·61. Examples.

(i) *Shew that all parallel frusta of a uniform cone of the same thickness exert equal attractions at the vertex.*

Since a cone of any form can be subdivided into a number of cones of small solid angle having the same vertex, and since the attraction of a thick frustum is got by adding the attraction of a number of thin ones, it is sufficient to prove the theorem for parallel thin slices of a cone of small solid angle.

Let PQ be a slice of the cone of small thickness t, and let ρ be the density, O the vertex, $OQ = r$ and θ the angle the normal to PQ makes with OQ. Then the area $PQ = r^2 \sec \theta \, d\omega$ **(1·4)**, and the mass of the frustum $= \rho r^2 t \sec \theta \, d\omega$.

The attraction at O of this frustum is therefore $\rho t \sec \theta \, d\omega$,

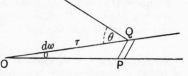

and this has the same value for all parallel frusta of the same thickness.

(ii) *Prove that the potential of a solid of uniform density ρ at an external point P can be represented by a surface integral $\frac{1}{2}\rho \int \cos \theta \, dS$ over the surface of the solid, where θ is the angle between the inward normal to dS and the line joining dS to P.*

Let a cone of small solid angle $d\omega$ and vertex P cut the surface of the solid in elements of area dS_1, dS_2 at A, B, where the inward normals make angles θ_1, θ_2 with the line BAP. Let

$$AP, \; BP \equiv r_1, r_2.$$

The mass of an element of volume of the cone at a distance r from P is, as in 1·42, $\rho r^2 \, d\omega \, dr$. Hence the mass of the cone between A and B produces at P a potential

$$\int \rho \frac{r^2 \, d\omega \, dr}{r} = \tfrac{1}{2}\rho \, (r_2{}^2 - r_1{}^2) \, d\omega$$
$$= \tfrac{1}{2}\rho \, (dS_2 \cos \theta_2 + dS_1 \cos \theta_1).$$

If we take the sum for all cones which intersect the solid we shall get $\tfrac{1}{2}\rho \int \cos \theta \, dS$ integrated over the surface.

EXAMPLES

1. A sphere of 150 kilogrammes, placed with its centre 30 cm. vertically below that of another sphere, is found to cause an apparent increase of the latter's weight by the $1·14/10^8$ part. What value does this imply for the constant of gravitation? [M. T. 1918]

2. Three equal uniform rods, each of length $2a$ and of mass m per unit length, form an equilateral triangle. Shew that the attraction on a particle of unit mass at a point on the straight line through the

centre of the triangle at right angles to its plane, and at a distance z from it, is

$$\frac{6maz}{\left(z^2+\frac{1}{3}a^2\right)\left(z^2+\frac{4a^2}{3}\right)^{\frac{1}{2}}}.$$ [C. 1906]

3. Prove that any plane closed rectilinear polygon of thin uniform attracting wire, all of whose sides touch one and the same circle, exerts no attraction at the centre of the circle.

[London Univ. 1927]

4. A uniform wire of mass M has the form of a semicircular arc of radius a together with the diameter of the semicircle. Find the attraction of the wire upon a particle of unit mass placed at the other end of the diameter through the middle point of the arc of the semi-circle. [London Univ. 1933]

5. Prove that the attraction of a thin uniform cylindrical shell of radius a and length l at a point on its axis at a distance b from one end and $l-b$ from the other ($b < \frac{1}{2}l$) is

$$\gamma\frac{M}{l}\left[\frac{1}{\sqrt{(a^2+b^2)}}-\frac{1}{\sqrt{\{a^2+(l-b)^2\}}}\right],$$

where M is the mass of the shell.

[The ends of the shell are open and circular.]

[London Univ. 1926]

6. Shew that the attraction of a uniform solid hemisphere at the centre of its plane base is $\frac{3}{2}\gamma M/a^2$, where M is its mass and a is its radius.

7. Prove that the potential of a circle of uniform surface density σ and radius a at a point on its circumference is $4\gamma\sigma a$.

[London Univ. 1932]

8. Prove that the potential of a plane annulus bounded by concentric circles at a point on its axis is $2\pi\gamma\rho\,(l'-l)$, where ρ is the mass per unit area and l, l' are the distances of the point from the inner and outer edges.

9. A uniform circular lamina is divided into two segments by a chord which subtends an angle $2\alpha\,(<\pi)$ at the centre. Prove that the attraction at the centre of the circle due to the minor segment is

$$2\gamma\sigma\left[\log\tan\left(\tfrac{1}{4}\pi+\tfrac{1}{2}\alpha\right)-\sin\alpha\right],$$

σ being the surface density of the lamina. [London Univ. 1931]

10. Prove that, if the surface density of a circular disc at any point distant r from the centre is λr, then the attraction at a point on the axis at a distance d is

$$\pi\gamma\lambda d\left\{\log\frac{\sqrt{(a^2+d^2)}+a}{\sqrt{(a^2+d^2)}-a}-\frac{2a}{\sqrt{(a^2+d^2)}}\right\}.$$

[London Univ. 1932]

11. Four uniform rods, each of the same mass per unit length, form a square $ABCD$. Any point P is taken inside the square and E, F, G, H are the feet of the perpendiculars from P upon AB, BC, CD, DA respectively. Prove that the potential at P due to the square $ABCD$ is the same as that due to four rods EF, FG, GH, HE each of the same mass per unit length as that of the rods forming the square.

[London Univ. 1935]

12. Two uniform rods AB and CD, of length a and mass m per unit length, are placed so that CD lies along a right bisector of AB, and C and D are distant a and $2a$ respectively from the rod AB. Shew that the attraction of either rod on the other is of magnitude

$$2\gamma m^2 \{\log_e (1 + \sqrt 5) + \log_e 2 - \log_e (1 + \sqrt{17})\}/a^2.$$

[London Univ. 1938]

13. Find the attraction of a uniform circular ring, of mass M and radius a, upon a thin uniform rod, of mass M' and length $2l$, lying along the axis of the ring with its middle point at a distance $x\,(<l)$ from the centre of the ring.

If the ring is fixed and the rod released from rest when its centre is at a *small* distance x from the centre of the ring and moves under the mutual attraction only, shew that the period of small oscillation is

$$2\pi(a^2 + l^2)^{\frac{3}{4}}/(\gamma M)^{\frac{1}{2}}.\qquad \text{[London Univ. 1935]}$$

14. Find the attraction of a uniform thin shell, in the form of the portion of a circular cylinder between two planes perpendicular to the axis, on a particle at the axis; and prove that a particle slightly displaced on the axis from the position of equilibrium will perform small oscillations of period $2\pi M^{-\frac{1}{2}} (a^2 + c^2)^{\frac{3}{4}}$, where $2a$ is the length and c the radius of the cylinder, and M its mass.   [C. 1892]

15. Prove that the potential due to a thin rod AB, of uniform line density m, at any point P in the line AB produced is $\gamma m \log (PA/PB)$. Two such rods, each of length $2a$, are placed so that they lie in the same straight line with their nearer ends at a distance $2a$ apart. Prove that a particle, released from rest at a point in the line of the rods at a small distance x from the point midway between them, begins to move with acceleration $16\gamma mx/9a^2$, approximately.

[London Univ. 1936]

16. Prove that the resultant attraction at P of a straight rod AB of uniform line density m is $2\gamma me/PG$ directed along PG, where e is the eccentricity of the ellipse of foci A, B passing through P, and G is the point in which the normal to the ellipse at P cuts AB.

17. The ends of a thin uniform rod are at the foci of a hyperbola, and a similar rod is placed parallel to it with its ends on the curve. Shew that the attraction between the two rods is inversely proportional to their distance apart.   [C. 1908]

18. Shew that the potential of a *closed* hemispherical shell, made of uniform thin material of surface density σ, at a point on the axis of symmetry at a distance a from its plane face remote from its curved surface is $2\pi\gamma\sigma a$, where a is the radius of hemisphere.

[London Univ. 1937]

19. AB, $A'B'$ are two parallel rods of line densities m, m' and are placed with the line joining their middle points perpendicular to the rods and of length c. Shew that the attraction between the rods is
$$\frac{2mm'}{c}(AB'-AA').$$ [C. 1914]

20. Two uniform rods AB, AC, each of mass m and length a, are freely jointed at A. Shew that the couple required to keep them at right angles against their mutual attraction is $(2-\sqrt{2})\gamma m^2/a$.

[London Univ.]

21. Two uniform thin rods AB and CD are of linear density λ and λ' and length $2a$ and $2c$ respectively. The midpoint E of AB and the midpoint F of CD are at a distance b apart, and AB, CD and EF are mutually perpendicular. Shew that the attraction between the rods is
$$4\gamma\lambda\lambda'\sin^{-1}\frac{ac}{\{(a^2+b^2)(b^2+c^2)\}^{\frac{1}{2}}},$$
where γ is the gravitational constant. [P. 1934]

22. Shew that the attraction of a uniform solid circular cylinder, of radius a and density ρ, at an external point P on the axis of the cylinder is
$$2\pi\gamma\rho a(\tan\tfrac{1}{2}\alpha-\tan\tfrac{1}{2}\beta),$$
where α, β are the angles subtended at P by the radii of the nearer and farther ends of the cylinder. [London Univ. 1925]

23. A finite solid, of uniform density ρ, is bounded by the paraboloid $x^2+y^2=4az$ and the plane $z=a$. Prove that the intensity of attraction at the origin is
$$2\pi\gamma\rho a[1-\sqrt{5}+4\log_e\tfrac{1}{2}(1+\sqrt{5})].$$

[London Univ. 1938]

24. A quantity of homogeneous matter of density ρ is in the form of a portion of a paraboloid of revolution bounded by a plane perpendicular to the axis. Prove that its attraction at its vertex tends, as c the length of the axis increases, to the limit $4\pi\rho a\log c/ae$, where $4a$ is the length of the latus rectum of the generating parabola.

25. A uniform solid, of density ρ, is in the form of the finite portion of a paraboloid of revolution of latus rectum $4a$ cut off by a plane perpendicular to the axis at a distance $2a$ from the vertex. Prove that the attraction at the centre of the circular end is
$$4\pi\gamma a\{\sqrt{2}-\log(1+\sqrt{2})\}.$$

[London Univ. 1933]

26. Shew that for a long rod of length $2l$ and line density m, the component attractions parallel and perpendicular to the rod are approximately

$$\frac{2\gamma m x}{l^2}\left(1+\frac{2x^2-3y^2}{2l^2}\right) \quad \text{and} \quad \frac{2\gamma m}{y}\left(1-\frac{y^2}{2l^2}\right),$$

where the origin is at the middle point of the rod and the axis of x along it, and terms of order equal to or higher than $(d/l)^4$ are neglected, where d is the distance from the middle point.   [M. T. 1919]

---

27. A homogeneous prism, infinite in length, whose cross-section is an equilateral triangle ABC attracts a particle at A; prove that the resultant attraction is $4\pi M/3a$, where M is the mass of a unit length of the prism and a is the length of a side of the triangle.   [C. 1908]

28. ABC is a normal cross-section of a uniform triangular prism of infinite length and of mass m per unit length. Shew that the resultant attraction at the point A is compounded of $4\gamma mA/a$ perpendicular to BC, and $4\gamma m \log(b/c)/a$ parallel to BC, the angle A being measured in radians.   [London Univ. 1925]

29. Two infinitely long uniform circular cylinders, each of mass m per unit length, are placed with their axes parallel. Prove that the potential, V, at a point, outside both cylinders, whose distances from their axes are r, r', is given by

$$V = \text{const.} - 2\gamma m \log(rr').$$

Further, prove that the lines of force outside the cylinders are arcs of rectangular hyperbolas.   [London Univ. 1936]

30. A tetrahedron is made from a uniform sheet of thin metal. Shew that, if the law of attraction were the inverse cube, a particle would be in equilibrium if placed at the centre of the inscribed sphere.

31. Two thin straight uniform rods AB, CD are pivoted together at their middle points. Prove that the attraction between them reduces to a couple of moment $2mm'(AC \sim BC)\operatorname{cosec}\alpha$, where m and m' are their line densities and α is the angle between them.   [C. 1903]

32. The cross-section of a long uniform solid cylinder is a semicircle. Shew that the direction of the resultant attraction at a point on an edge far from the ends of the cylinder makes an angle $\tan^{-1}(2/\pi)$ with the plane face.   [C. 1928]

33. Shew that a uniform thin square lamina of mass M, whose side is of length l, exerts an attraction at a point P on the perpendicular to the lamina through one corner, whose component parallel to the lamina is

$$\frac{\sqrt{2}\gamma M}{l^2}\log\left[\frac{(h^2+l^2)^{\frac{1}{2}}\{l+(h^2+l^2)^{\frac{1}{2}}\}}{h\{l+(h^2+2l^2)^{\frac{1}{2}}\}}\right],$$

where h is the distance from P to the lamina.   [London Univ. 1931]

34. A plane sheet of matter of uniform surface density is bounded by two infinite parallel straight edges and extends to infinity in both directions. Find the components of the attraction parallel to and at right angles to its plane; and shew that the loci of the points for which these are constant are two sets of orthogonal cylinders. [C. 1898]

35. Shew that, if a uniform straight rod CD were free to slide in a smooth groove parallel to a fixed similar rod AB, its time of oscillation about its equilibrium position would be

$$2\pi \left\{ \frac{2\gamma m}{CD} \left(\frac{1}{BD} \sim \frac{1}{AD} \right) \right\}^{-\frac{1}{2}},$$

where m is the mass of unit length of the rod AB, and AD, BD are distances in the equilibrium position. [C. 1926]

36. Determine the potential of a solid uniform right circular cone at the centre of its base. [London Univ. 1931]

37. A vertical solid right circular cylinder of height h, radius a and density ρ, bounded by plane ends perpendicular to the axis, is divided by a plane through the axis into two parts. Shew that the horizontal attraction of either part at the centre of the base is

$$2\rho h \log \left[\{a + \sqrt{(a^2 + h^2)}\}/h \right].$$

38. Shew that two finite surfaces having the same surface density, which are geometrically inverse to one another, exert equal attractions at the centre of inversion. [C. 1906]

39. If the law of attraction of a thin uniform straight rod is that of the inverse cube, shew that the equipotential surfaces are generated by the revolution of a family of curves, whose polar equation is

$$(r^4 + a^4 - 2r^2 a^2 \cos 2\theta)^{\frac{1}{2}} = 2ar \sin \theta \cdot \operatorname{cosec} (cr \sin \theta)$$

about the initial line, $2a$ being the length of the rod and c a parameter. [C. 1907]

40. The mass per unit length at a point P of a rod ACB, of which A and B are the ends and C a particular point, varies as $CP/\sqrt{(AP \cdot PB)}$. Prove that the resultant attraction of the rod at any point at which AC, CB subtend equal angles acts through C. [M. T. 1902]

41. An infinitely long thin rod, of uniform line density m, and a thin uniform rod of mass M and finite length $2a$ are so placed that the shortest distance between them is of length b and bisects the finite rod. Prove that the resultant attraction of either rod on the other is equivalent to a wrench of intensity

$$2\gamma Mm\phi/(a \sin \theta)$$

and pitch $\quad b \cot \theta (\tan \phi - \phi)/\phi,$

where θ is the angle between the rods and $\phi = \tan^{-1} [(a \sin \theta)/b]$. [London Univ. 1938]

42. A solid uniform hemisphere is of mass M and radius a. Prove that its attraction at any point P of its axis on the convex side of the curved surface is equal to

$$\frac{M}{r^2}\left(1 + \frac{R^3 + 2r^3 - 3r^2 R}{a^3}\right),$$

where r and R are the distances of P from the centre and rim of the base. [C. 1902]

43. Shew that, if a denotes the radius and ρ the density of a hemispherical hill and R the radius and σ the mean density of the earth, the difference of the latitudes as measured by a plumb line of stations on the north and south sides of the base of the hill is increased by the attraction of the hill by approximately $a\rho/R\sigma$.

44. Consider the effect on the plumb line of a deep crevasse running east and west, and shew that if the density of the earth's crust is greater than two-thirds of the earth's mean density the north edge of the crevasse will have a lower apparent latitude than the south edge.

45. Prove that at the edge of a very long straight depression of rectangular section, whose depth h is small compared with its breadth c, the plumb line will be deflected through an angle approximately equal to $2\mu h(1 + \log c/h)/g$ from its mean position, where μ is the attraction at unit distance of unit volume of the superficial strata supposed condensed to a point. [M. T. 1887]

46. The surface density of a circular disc of radius a is $\sigma_0(1 - r^2/a^2)^{\frac{1}{2}}$ at a distance r from the centre of the disc. Prove that the potential at a point on the axis of the disc at a distance $a \cot \phi$ from its centre is

$$\pi\gamma\sigma_0 a\,(\phi\operatorname{cosec}^2\phi - \cot\phi).$$

[London Univ. 1927]

47. Prove that a lamina, of uniform density σ in the form of the segment of a circle, attracts a point P on the remaining arc of the circle with a force tending towards D whose magnitude is

$$\frac{\sigma}{a}\left\{DP\log\frac{DP + DA}{DP - DA} - 2DA\right\},$$

where AB is the base of the segment, D the middle point of its arc and a the radius of the circle. [C. 1906]

48. Prove that a solid regular tetrahedron of uniform density ρ exerts at a vertex an attraction of intensity

$$\gamma\rho h\,(3\cos^{-1}\tfrac{1}{3} - \pi),$$

where h is the perpendicular distance of a vertex from the opposite face. [London Univ. 1936]

49. Prove that a thin triangular lamina ABC of uniform surface density σ exerts, at a point on the line through A perpendicular to the plane of the triangle and at a distance k from A, an attraction the intensity of whose component perpendicular to the lamina is

$$\gamma\sigma\left[A - \sin^{-1}\frac{k\cos B}{\surd(k^2+p^2)} - \sin^{-1}\frac{k\cos C}{\surd(k^2+p^2)}\right],$$

where p is the perpendicular distance of A from BC.

[London Univ. 1939]

50. If a uniform lamina is the portion of the plane xy containing the origin and bounded by the hyperbola $x^2/a^2 - y^2/b^2 = 1$, shew that the z component of its attraction at any point on the ellipse $x^2/(a^2+b^2) + z^2/b^2 = 1$, $y = 0$ is $\dfrac{2\pi mah}{\surd(a^2h^2+b^4)}$, where h is the z co-ordinate of the point.

[C. 1900]

Chapter III

ATTRACTION AND POTENTIAL AT INTERNAL POINTS. SPHERES

3·1. Attraction and Potential at internal points. In the previous chapter we confined our attention to the attraction and potential at points external to the attracting matter. We have now to consider what meaning to attach to the same terms at points inside the attracting matter. We have no reason to suppose that the law of inverse squares holds good at small distances comparable with the size of the molecules, and we have explained in **2·3** what, for mathematical convenience, we understand by the term 'continuous body'. Our definitions of attraction and potential at an external point imply the existence of a separate attracted particle at the point under consideration. Such a particle cannot exist inside a continuous body because two particles cannot occupy the same space simultaneously. We therefore imagine that there is a small cavity in the body surrounding the attracted particle placed at the chosen point. We assume that we can calculate the attraction and potential at this point by our former rules, since the attracted particle is not in contact with the matter. Then we define the attraction and potential at the same point in the continuous body to be the limits to which the attraction and potential at the point in the cavity tend as the cavity decreases in size and ultimately vanishes.

3·11. Convergent integrals. By reference to the test given in **1·8** for the convergence of volume integrals of the form $\int \frac{\rho \, dv}{r^n}$, namely that the integral converges at points within the region of integration when $n < 3$, we see that the integral $\int \frac{\rho \, dv}{r}$, which represents the potential, converges at points within the attracting matter; and the same is true of the integrals of the form $\int \dfrac{\rho\,(x-x')}{r^3}\,dv$ which represent the attraction components, since $\left|\dfrac{x-x'}{r^3}\right|$ is of order $\dfrac{1}{r^2}$. It may be shewn, however, that when the attracting matter consists of a surface distribution the integrals which represent tangential components of attraction at a point in the matter are not convergent in the sense of 1·8 but depend on the shape of the vanishing cavity.

3·2. Uniform thin spherical shell. Let m be the mass per unit area of a thin spherical shell of radius a and centre O.

Attraction at an internal point. With any internal point P as vertex construct a cone of small solid angle $d\omega$ intersecting the surface in elements QQ', RR' of areas dS, dS'. The attractions at P of these elements are $m\,dS/QP^2$ and $m\,dS'/RP^2$ in opposite directions. But

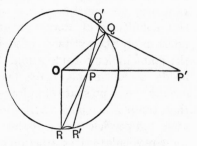

$$dS = QP^2 \sec OQP\,d\omega,$$

and

$$dS' = RP^2 \sec ORP\,d\omega,$$

and the angles OQP, ORP are equal, so that the elements QQ', RR' of the shell exert equal and opposite attractions at P. Since the whole shell can be divided into similar pairs of elements by taking cones in all directions round P, it follows that the resultant attraction of the shell at P is zero.*

Attraction at an external point. Let P' be any external point and P its inverse, so that $OP.OP' = a^2$. The resultant attraction at P' is by symmetry along $P'O$; and with the same construction, the element QQ' exerts an attraction $m\,dS/QP'^2$ along $P'Q$, and resolving this in direction $P'O$ we get

$$\frac{mQP^2 \sec OQP\,d\omega}{QP'^2}\cos OP'Q.$$

* This proof was given by Newton, *Principia*, Proposition LXX.

But by similar triangles $QP : QP' = OQ : OP'$, and the angles OQP, $OP'Q$ are equal. Therefore the element QQ' contributes an amount $ma^2 d\omega / OP'^2$ to the resultant attraction. By taking cones in all directions round P, we get for the attraction at P' of the whole shell $4\pi ma^2 / OP'^2$, or M/OP'^2, where M denotes the mass of the shell. It follows that the attraction of the shell at external points is the same as if its mass were collected into a particle at its centre. This result is also due to Newton.[*]

3·21. *Potential at an internal point.* Since the attraction is zero throughout the interior of the shell, there can be no variation in the potential, or the potential is constant. The potential at every point in the interior is therefore the same as the potential at the centre, i.e. M/a, where M denotes the whole mass, since every element of M is at the same distance a from O.

This result may also be obtained directly without assuming a knowledge of the attraction, thus: The contribution to the potential at P of the elements QQ', RR' is

$$m\,dS/QP + m\,dS'/RP$$
$$= mQP \sec OQP\,d\omega + mRP \sec ORP\,d\omega,$$

and since the angles OQP, ORP are equal, this

$$= mQR \sec OQP\,d\omega = 2ma\,d\omega.$$

If we take all such double cones round P we take the shell twice over, we therefore put 2π for $d\omega$, and get for the whole potential $4\pi ma$, or M/a.

Potential at an external point. Let $OP' = r$, then since the force at distance r is M/r^2 in the direction in which r decreases, therefore

$$\frac{dV}{dr} = -\frac{M}{r^2}$$

and

$$V = \frac{M}{r} + C.$$

But the potential vanishes at an infinite distance, therefore $C = 0$ and $V = M/r$.

[*] *Principia*, Proposition LXXI.

This result may also be obtained independently, thus: The contribution of the elements QQ', RR' to the potential at P'

$$= m\,dS/QP' + m\,dS'/RP'$$

$$= \frac{mQP^2 \sec OQP \, d\omega}{QP'} + \frac{mRP^2 \sec ORP \, d\omega}{RP'}$$

$$= m\,d\omega \left(\frac{aQP}{OP'} \sec OQP + \frac{aRP}{OP'} \sec ORP \right), \text{ by similar triangles,}$$

$$= \frac{ma\,d\omega}{OP'} . QR \sec OQP = \frac{2ma^2 d\omega}{OP'},$$

and, as before, for the contributions of the whole shell we put 2π for $d\omega$, and obtain

$$V = \frac{4\pi ma^2}{OP'} = \frac{M}{r}.$$

3·3. Attraction of a uniform solid sphere. Such a sphere may be regarded as composed of a series of concentric thin spherical shells, and the required results may be obtained by summation. Let a be the radius and ρ the density of the sphere.

Attraction at an internal point. Take a point P at a distance r from the centre. Imagine a thin shell of matter of external and internal radii $r + \epsilon$ and $r - \epsilon$ to be removed and consider the at- traction at a point P in this cavity. By **3·2** the concentric shells external to the cavity exert no attraction at P, and those internal to the cavity attract as though their masses were concentrated at the centre O. Hence the attraction at P is the limit as $\epsilon \to 0$ of $\frac{4}{3}\pi\rho (r - \epsilon)^3/r^2 = \frac{4}{3}\pi\rho r$; i.e. the attraction of a uniform solid

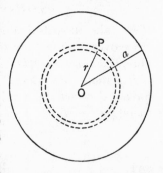

sphere at an internal point is directly proportional to the distance from the centre.

Attraction at an external point. Since each of the concentric shells attracts at an external point as though its mass were collected at its centre, the same is true of the solid sphere, and its attraction is represented by M/r^2, where M is its mass and r the distance of an external point from the centre.

3·31. Potential of a uniform solid sphere.

At an internal point. Adopting the method of **3·3**, let R denote the radius of a shell external to the cavity. Its mass is $4\pi\rho R^2 dR$, so that by **3·21** the potential it produces at a point P inside itself is $4\pi\rho R dR$, and the potential at P due to all such external shells is given by

$$\int_{r+\epsilon}^{a} 4\pi\rho R \, dR = 2\pi\rho \left\{ a^2 - (r+\epsilon)^2 \right\}.$$

Also, by **3·21**, the shells of radius less than that of the cavity produce the same potential as if the mass were collected at O, i.e. $\frac{4}{3}\pi\rho\,(r-\epsilon)^3/(r-\epsilon)$. Hence the whole potential at P is the limit as $\epsilon \to 0$ of

$$\tfrac{4}{3}\pi\rho\left\{(r-\epsilon)^2 + \tfrac{3}{2}\left[a^2 - (r+\epsilon)^2\right]\right\} = \tfrac{2}{3}\pi\rho\,(3a^2 - r^2) \quad \dots(1).$$

At an external point. Since each of the concentric shells produces at an external point a potential equal to its mass divided by the distance of the point from the centre, the same is true for the solid sphere.

To deduce the potential from the attraction. At an external point, as in **3·21**, we have $\dfrac{dV}{dr} = -\dfrac{M}{r^2}$, leading as before to

$$V = M/r \qquad \dots\dots\dots\dots\dots\dots(2).$$

At an internal point, since the attraction is $\frac{4}{3}\pi\rho r$, we have

$$\frac{dV}{dr} = -\tfrac{4}{3}\pi\rho r,$$

so that $\qquad\qquad V = C - \tfrac{2}{3}\pi\rho r^2 \qquad \dots\dots\dots\dots\dots(3),$

where the constant may be determined from the fact that the potential at the surface $r=a$ must be the same whether obtained by putting $r=a$ in the formula that gives the internal or the external potential. But from (2), when $r=a$, $V=\frac{4}{3}\pi\rho a^2$ and substituting in (3) gives $C=2\pi\rho a^2$, and

$$V=\tfrac{2}{3}\pi\rho\,(3a^2-r^2),$$

as before.

It is evident that the results obtained for the external field of force of a uniform shell bounded by concentric spheres, or for a uniform solid sphere, are also true for heterogeneous shells or solids provided they can be divided up into thin shells or strata of uniform density.

3·4. *When the potential of any distribution of matter on the surface of a sphere is known at internal points it can be found at external points.* Let P, P' be inverse points with regard to the sphere, so that, if a is the radius, $OP.OP'=a^2$. Then, if Q be any point on the surface, by similar triangles,

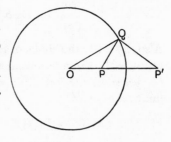

$$QP:QP'=OQ:OP'=a:OP'.$$

But, if m denotes the mass of an element at Q, the potentials it produces at P, P' are m/PQ and $m/P'Q$ and these are in the constant ratio $OP':a$ for all positions of Q. Hence, if V, V' denote the potentials at P, P' of the distribution of matter on the sphere, we have $V'=\dfrac{a}{OP'}\,V$.

For example, the potential at an internal point of a uniform thin shell of mass M and radius a is given by $V=M/a$ (**3·21**), therefore the potential at an external point at a distance r from the centre is given by $V'=M/r$.

3·5. *Attraction at an internal point of a heterogeneous sphere in which the density is a function of the distance from the centre.* Let a be the radius and let $f(R)$ denote the density at a distance R from the centre. Consider the attraction at a point P whose distance from the centre is r. As in **3·2**, shells of radius greater than r exert no attraction at P and shells of radius less than r

attract with a force Mass$/r^2$. Hence it is easy to see that the required attraction is

$$\frac{1}{r^2}\int_0^r 4\pi f(R)\,R^2\,dR.$$

3·51. Example. *Find the law of density when the attraction at an internal point of a sphere of given mass is constant and equal to its value at the surface.*

Let a be the radius and ρ the mean density so that the mass is $\frac{4}{3}\pi\rho a^3$. Then with the notation of **3·5**

$$\frac{1}{r^2}\int_0^r 4\pi f(R)\,R^2\,dR=\tfrac{4}{3}\pi\rho a,$$

so that $$\int_0^r f(R)\,R^2\,dR=\tfrac{1}{3}\rho a r^2.$$

Differentiating with regard to r, we get
$$f(r)\,r^2=\tfrac{2}{3}\rho a r,$$
and $f(r)=\dfrac{2}{3}\dfrac{\rho a}{r}$ gives the law of density.

3·52. Uniform shell bounded by non-intersecting spheres. When the boundary spheres are not concentric, let A, B be their centres and a, b their radii. Let ρ be the density of the matter between the spheres, and let r, r' be the distances of *any* point P from A and B. The field of force due to the given shell is the difference of the fields due to two uniform spheres of radii a, b and centres A, B. Hence, using the results of **3·31**, the potential has the following values:

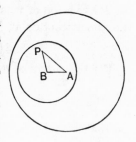

outside the shell $V_1=\tfrac{4}{3}\pi\rho\left(\dfrac{a^3}{r}-\dfrac{b^3}{r'}\right)$;

in the substance $V_2=\tfrac{2}{3}\pi\rho\left(3a^2-r^2-\dfrac{2b^3}{r'}\right)$;

and in the hollow $V_3=\tfrac{2}{3}\pi\rho\,(3a^2-r^2-3b^2+r'^2).$

Also, using the results of **3·3**, the resultant attraction has components as follows: when P is outside the shell $\tfrac{4}{3}\pi\rho\,\dfrac{a^3}{r^2}$

along PA and $\frac{4}{3}\pi\rho\dfrac{b^3}{r'^2}$ along BP; in the substance $\frac{4}{3}\pi\rho r$ along

PA and $\frac{4}{3}\pi\rho\dfrac{b^3}{r'^2}$ along BP; and in the hollow $\frac{4}{3}\pi\rho r$ along PA

and $\frac{4}{3}\pi\rho r'$ along BP. And in this last case the forces compound into a resultant force $\frac{4}{3}\pi\rho BA$ parallel to BA; so that the force inside such a hollow shell is constant in magnitude and direction, i.e. *a uniform field of force.*

3·6. Long uniform solid cylinder. We may regard a solid cylinder as composed of thin cylindrical shells having the same axis. We have seen in **2·42** that the attraction of a uniform long thin circular cylindrical shell is zero at internal points, and at external points is the same as if the mass were concentrated uniformly along the axis of the cylinder. For the attraction at an internal point P, all the shells external to P exert no attraction at P, and the resultant attraction at P is therefore due to the shells nearer to the axis than P, so that if M be the mass of unit length of the cylinder and a its radius, and r the distance of P from the axis, the cylinder attracts at P like a rod of mass Mr^2/a^2 per unit length along the axis; i.e. the attraction at P is

$$2\,(Mr^2/a^2)/r = 2Mr/a^2.$$

For the potential we have

$$\frac{dV}{dr} = -\frac{2Mr}{a^2},$$

so that $\qquad\qquad V = \text{const.} - Mr^2/a^2,$

where $r < a$. But from **2·42** the potential at an external point is

$$V = 2M\log 2l - 2M\log r,$$

where $2l$ is the length of the cylinder. And in order that these expressions may give the same value for V on the surface $r = a$ (**3·7**), the internal potential must be

$$V = 2M\log 2l - 2M\log a + M\,(1 - r^2/a^2).$$

3·7. *Variation in the attraction in crossing a surface on which there exists a thin layer of attracting matter.* Let P_1, P_2 be two points on opposite sides of the surface close to one another on the same normal to the surface. Let m be the mass per unit area of a small circular disc of the surface between P_1 and P_2.

The points may be regarded as so close to the disc that, in accordance with the result of 2·5, it exerts equal attractions $2\pi m$ at P_1 and P_2 both directed towards the surface.

The attraction of the *rest of the matter* on the surface (or elsewhere) may be represented in the neighbourhood of P_1 and P_2 by a normal component X taken in the sense $P_1 P_2$ and a component Y in some direction tangential to the surface. And the values of these components at P_1, P_2 will differ by infinitesimals of the order $P_1 P_2$. For if X be the value at P_1 and $P_1 P_2 = \delta n$, the value at P_2 is $X + \dfrac{\partial X}{\partial n}\delta n$ + higher powers, and the infinitesimal may be neglected if we neglect the thickness of the layer of matter.

We now compound the attraction of the small disc and the attraction of the rest of the matter, and let X_1, Y_1 and X_2, Y_2 denote the components of the whole attraction at P_1, P_2, taking X_1, X_2 in the sense $P_1 P_2$, and we have

$$X_1 = X + 2\pi m, \quad Y_1 = Y$$

and

$$X_2 = X - 2\pi m, \quad Y_2 = Y.$$

Therefore $$X_2 - X_1 = -4\pi m, \quad Y_2 = Y_1 \quad \ldots\ldots\ldots\ldots(1),$$

or the tangential component of attraction is unaltered and the normal component is diminished by $4\pi m$ in crossing the surface.

As regards the potential, it is clear from 2·51 that a thin disc of matter produces the same potential at points on opposite sides close to itself, and the potentials of the rest of the matter at P_1, P_2 only differ by an infinitesimal, which may be neglected if we neglect the thickness of the layer of matter. We conclude that the potential is continuous across such a surface as we are considering, in the sense that there is no jump in its numerical value. This is illustrated by the case of the thin spherical shell of 3·21. The internal and external potentials are represented by different functions, but they have the same value on the sphere $r = a$.

Reverting now to the result (1), if we use V_1, V_2 to denote the potential functions on opposite sides of the surface, and dn to denote an element of the normal taken positively from P_1 to P_2, we have, at the surface,

$$\left. \begin{aligned} V_1 &= V_2 \\ \text{and} \qquad \frac{\partial V_2}{\partial n} - \frac{\partial V_1}{\partial n} &= -4\pi m \end{aligned} \right\} \quad \dots\dots\dots\dots(2).$$

We remark that m denotes the mass per unit area or the 'surface-density' of the layer, and the thickness of the layer is neglected; this implies that a finite mass is condensed into a volume of no thickness, so that the *volume density* of the matter is infinite.

3·71. Attraction of a thin uniform spherical shell at a point of itself. The attraction at a point of itself of a thin layer of matter depends on the shape of the gap in the surface in which the attracted particle is placed. We may define the *principal value* of the attraction as the limiting value of the attraction at the centre of a circular hole when its radius tends to zero. We proceed to calculate this for a spherical shell. Let m be the mass per unit area of the shell, and omitting a small circular element of the shell surrounding a point P, consider the attraction of the rest of the shell at P. Let an element QQ' of area dS subtend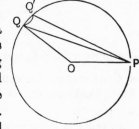
a solid angle $d\omega$ at P. The attraction of this element at P is $m\,dS/PQ^2$ along PQ, and resolving this along PO, which is clearly the direction of the resultant attraction, we get $m\,dS \cos OPQ/PQ^2$, or $m\,dS \cos OQP/PQ^2$, which is equal to $m\,d\omega$. If now we allow the gap in the shell round P to shrink to vanishing point, we see that we have to take the sum

$\Sigma m \, d\omega$ for all cones on one side of the tangent plane at P, so that the resultant attraction is $2\pi m$.

3·8. Mutual work of two attracting systems.

Let there be a system of particles of masses $m_1, m_2, m_3, \ldots$ at the points $A_1, A_2, A_3, \ldots$ and a second system of masses $m_1', m_2', m_3', \ldots$ at the points $B_1, B_2, B_3, \ldots$.

If we suppose that initially the particles of the second system are scattered at an infinite distance and are brought from thence up to their final positions $B_1, B_2, B_3, \ldots$ by the attractions of the first system, the work done by these attractions is represented by

$$V_1 m_1' + V_2 m_2' + V_3 m_3' + \ldots \quad \text{or} \quad \Sigma V m',$$

where $V_1, V_2, V_3, \ldots$ are the potentials of the first system at $B_1, B_2, B_3, \ldots$.

Similarly the work done by the attractions of the second system in bringing the particles of the first system from a state of infinite diffusion up to their final positions $A_1, A_2, A_3, \ldots$ is represented by

$$V_1' m_1 + V_2' m_2 + V_3' m_3 + \ldots \quad \text{or} \quad \Sigma V' m,$$

where $V_1', V_2', V_3', \ldots$ are the potentials of the second system at $A_1, A_2, A_3, \ldots$.

If we denote the distances $A_1 B_1, \, A_1 B_2, \, \ldots \, A_p B_q, \, \ldots$ by $r_{11}, r_{12}, \ldots r_{pq}, \ldots$, so that

$$V_1 = \frac{m_1}{r_{11}} + \frac{m_2}{r_{21}} + \frac{m_3}{r_{31}} + \ldots,$$

$$V_2 = \frac{m_1}{r_{12}} + \frac{m_2}{r_{22}} + \frac{m_3}{r_{32}} + \ldots,$$

$$\ldots\ldots\ldots\ldots\ldots\ldots\ldots\ldots\ldots,$$

it is evident that each of the expressions $\Sigma V m'$, $\Sigma V' m$ is equal to the sum $\Sigma m m'/r$ extended to the product of masses of every pair of particles, one of each system, divided by their distance, so that

$$\Sigma V m' = \Sigma V' m \quad \ldots\ldots\ldots\ldots\ldots\ldots(1).$$

For continuous masses we have in like manner

$$\int V dm' = \int V' dm \quad \dots\dots\dots\dots\dots(2),$$

where dm, dm' denote elements of mass of the systems and V, V' the potentials of the first system at dm' and of the second system at dm.

These expressions may be called the **mutual work** of the two systems, or the **exhaustion of their potential energy**. For potential energy can be considered to exist in matter in a state of infinite diffusion, and work is done and potential energy lost when the matter assumes a more condensed form, whether under the attractions of another system or under the attractions of its component parts. When we speak of the *mutual potential energy* of two attracting systems we should therefore prefix a negative sign to either of the expressions (1) or (2).

3·81. Lost potential energy of a gravitating system. Let the system consist of particles of masses $m_1, m_2, m_3, \dots$ at points $A_1, A_2, A_3, \dots$. Let us find the work done by the mutual attractions of the particles in assuming these positions starting from a state of infinite diffusion. Let m_1 be in its final position A_1; the work done by the attraction of the particle m_1 on the particle m_2 as it comes from an infinite distance to A_2 is $m_1 m_2/r_{12}$. Then these two particles attract the third and do work $m_1 m_3/r_{13} + m_2 m_3/r_{23}$ in bringing m_3 to A_3 and so on; so that the total work done may be represented by $\Sigma m_p m_q/r_{pq}$, where the summation extends to every pair of particles.

Let
$$V_1 = \frac{m_2}{r_{12}} + \frac{m_3}{r_{13}} + \dots,$$

$$V_2 = \frac{m_1}{r_{21}} + \frac{m_3}{r_{23}} + \dots,$$

$$\dots\dots\dots\dots\dots$$

Then V_1 is the potential at A_1 of $m_2, m_3 \dots$; V_2 is the potential at A_2 of $m_1, m_3 \dots$; and so on.

It follows that the total work done is equal to $\frac{1}{2}\Sigma mV$, the factor $\frac{1}{2}$ being introduced because the sum ΣmV contains every term such as $m_1 m_2/r_{12}$ twice.

This expression $\frac{1}{2}\Sigma mV$ represents the work done by the mutual attractions of the system of particles or *the exhaustion of their potential energy* as they come together from a state of infinite diffusion.

If the system forms a continuous body, its lost potential energy can be found by dividing it into elements dm and taking the limit of $\frac{1}{2}\Sigma V\,dm$. Here V is the potential at the position of dm of the rest of the body, and its limit is, by definition, the potential of the body at the point to which dm contracts. Thus the lost potential energy is

$$\frac{1}{2}\int V\,dm,$$

where V is the potential of the body at a point of itself and dm is an element of the body at the point. The potential energy of the body may be denoted by the same expression with a negative sign.

It used to be supposed that the duration of the sun's heat might be calculated by assuming that, as the matter forming the sun condensed from a state of infinite diffusion, the mechanical potential energy lost was converted into heat, and then estimating how long it would take for this heat to be lost by radiation.

3·9. Examples. (i) *The density of a sphere varies as the depth below the surface; shew that the resultant attraction is greatest at a depth equal to $\frac{1}{3}$ of the radius, and that the value there is $\frac{4}{9}$ of the value at the surface.*

Let a be the radius, and let the density ρ at a distance r from the centre be given by $\rho = \lambda(a - r)$.

The attraction at a distance R from the centre, when $R < a$, is due to the sphere of radius R and its value is

$$F = \frac{1}{R^2}\int_0^R 4\pi\rho r^2\,dr = \frac{4\pi\lambda}{R^2}\int_0^R (a-r)\,r^2\,dr$$

$$= 4\pi\lambda\left(\frac{aR}{3} - \frac{R^2}{4}\right) = \pi\lambda\{\tfrac{4}{9}a^2 - (\tfrac{2}{3}a - R)^2\}.$$

The greatest value of F occurs where $R = \frac{2}{3}a$, i.e. at a depth equal to $\frac{1}{3}a$, and this greatest value is $\frac{4}{9}\pi\lambda a^2$; whereas the value at the surface $R = a$ is seen to be $\frac{1}{3}\pi\lambda a^2$.

(ii) *Prove that, if the law of potential be* $\dfrac{A}{r}e^{-r/\lambda}$, *the potential of a uniform thin spherical shell of radius a at an external point is the same as that of a particle of mass* $\sinh\dfrac{a}{\lambda}\bigg/\dfrac{a}{\lambda}$ *that of the shell placed at the centre.*

[M. T. 1918]

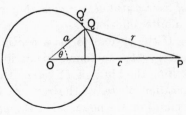

Let P be an external point at a distance c from the centre O.

Consider a narrow zone of the shell of breadth QQ' at a distance r from P. If θ denotes the angle QOP, the area of the zone is $2\pi a^2 \sin\theta\, d\theta$; so that if m is the mass of unit area, the potential at P due to this zone is

$$2\pi m a^2 \frac{A}{r}e^{-r/\lambda}\sin\theta\, d\theta,$$

and the whole potential is

$$V = 2\pi m a^2 A \int_0^\pi \frac{e^{-r/\lambda}}{r}\sin\theta\, d\theta.$$

But $r^2 = a^2 + c^2 - 2ac\cos\theta$, so that $r\, dr = ac\sin\theta\, d\theta$, and

$$V = 2\pi m \frac{a}{c} A \int_{c-a}^{c+a} e^{-r/\lambda}\, dr = 2\pi m \frac{a}{c} A\lambda\, (e^{-(c-a)/\lambda} - e^{-(c+a)/\lambda})$$

$$= 4\pi m \frac{a}{c} A\lambda e^{-c/\lambda}\sinh\frac{a}{\lambda}.$$

Also the potential at P of a mass $4\pi m a^2$ at O is, with the same law of potential, $4\pi m a^2 \dfrac{A}{c}e^{-c/\lambda}$; hence the result follows.

The reader will notice that this method of direct integration can be applied for finding the potential inside or outside the shell with the ordinary law of potential.

(iii) *Prove that the attraction of a uniform elliptic disc at a focus is* $2\pi m\{1 - \sqrt{(1-e^2)}\}/e$ *if the vanishing cavity round the focus is circular with the focus as centre, and zero if the cavity is bounded by a similar and similarly situated ellipse with the focus as centre of similarity.*

Let $l/r = 1 - e\cos\theta$ be the equation of the boundary ellipse and $r = a$ that of the small circular cavity round the focus.

The attraction is easily seen to be given by

$$\int_0^{2\pi}\int_a^{l/(1-e\cos\theta)} \frac{\cos\theta}{r^2} mr\, d\theta\, dr,$$

and integrating with regard to r, this

$$= m\int_0^{2\pi}\cos\theta\log\frac{l}{a\,(1-e\cos\theta)}\, d\theta.$$

Then, integrating by parts, this

$$= m \left[\sin\theta \log \frac{l}{a(1-e\cos\theta)} \right]_0^{2\pi} + m \int_0^{2\pi} \frac{e\sin^2\theta\, d\theta}{1-e\cos\theta}$$

$$= m \int_0^{2\pi} \left\{ \cos\theta + \frac{1}{e} - \frac{1-e^2}{e(1-e\cos\theta)} \right\} d\theta$$

$$= \frac{2\pi m}{e} \{1 - \surd(1-e^2)\}.$$

But if the inner boundary, which gives the lower limit of integration with regard to r, be $l'/r = 1 - e\cos\theta$, the integral of $\dfrac{\cos\theta}{r}$ with respect to r is $\cos\theta \log l/l'$, and integrating with respect to θ the result is zero.

(iv) *Prove that the gravitational potential energy of a thin uniform circular disc of mass m and radius a is $C - 8\gamma m^2/3\pi a$, where C is independent of a, and γ is the constant of gravitation.*

[M. T. 1913]

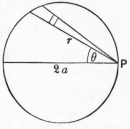

If σ denotes the surface density, the potential at any point P on the edge of such a disc is easily seen to be

$$\gamma\sigma \int_{-\frac{1}{2}\pi}^{\frac{1}{2}\pi} \int_0^{2a\cos\theta} \frac{r\,d\theta\,dr}{r}$$

$$= \gamma\sigma \int_{-\frac{1}{2}\pi}^{\frac{1}{2}\pi} 2a\cos\theta\, d\theta = 4\gamma\sigma a.$$

Now suppose that the particles which make up the disc are initially in a state of diffusion at an infinite distance and that C denotes their potential energy in this state. Let the particles be assembled in concentric rings. When the radius of the disc is r, the potential at its edge is, from above, $4\gamma\sigma r$, so that the work done in assembling a narrow ring of breadth dr is $4\gamma\sigma r \cdot 2\pi\sigma r\, dr$. It follows that the work done by its own attractions in bringing together the whole disc is

$$8\pi\gamma\sigma^2 \int_0^a r^2\, dr = \tfrac{8}{3}\pi\gamma\sigma^2 a^3.$$

But $m = \pi\sigma a^2$, so that the work done $= 8\gamma m^2/3\pi a$, and the potential energy remaining $= C - 8\gamma m^2/3\pi a$.

EXAMPLES

1. Taking the earth as a uniform solid sphere of 4000 miles radius, shew that a pendulum clock would lose 10·8 sec. per day if taken to the bottom of a mine 1 mile in depth.   [M. T. 1921]

2. A small smooth cylindrical hole is drilled through a solid uniform sphere. Shew that a particle set free at the surface will oscillate in a period independent of the direction of the hole.

3. The density of an infinite circular cylinder of radius a at a distance r from its axis is proportional to $(a^2 - r^2)^{\frac{1}{2}}$. Shew that the attraction at an internal point is

$$\frac{2\gamma M}{a^3 r}\{a^3 - (a^2 - r^2)^{\frac{3}{2}}\},$$

where M is the mass of unit length of the cylinder, and γ is the gravitational constant.   [London Univ. 1926]

4. A sphere is described on a radius of another sphere as diameter. If the latter be a homogeneous sphere of mass M and radius a, shew that the resultant attraction on the portion included in the smaller sphere is $\frac{1}{16}M^2/a^2$.   [C. 1914]

5. A solid sphere of radius a is such that its density at any point is proportional to the nth power of the distance of that point from the centre of the sphere. Find the potential at any external point, and shew that the potential at an internal point at a distance r from the centre is $\dfrac{\gamma M}{n+2}\left(\dfrac{n+3}{a} - \dfrac{r^{n+2}}{a^{n+3}}\right)$, where M is the total mass of the sphere.

[London Univ. 1931]

6. A uniform solid sphere of mass M is cut in two by a diametral plane; shew that the resultant attraction between the halves is $\frac{3}{16}\gamma M^2 a^{-2}$, where a is the radius of the sphere, and γ the constant of gravitation.   [C. 1906]

7. Prove that the force required to separate the two parts of a solid uniform sphere divided in any manner is $\gamma MM' \cdot GG'/a^3$, where M, M' are the masses, and G, G' the centres of gravity of the two parts and a is the radius of the sphere.   [C. 1905]

8. P is a point on a uniform sphere of mass M and radius a. The sphere is divided into two parts by a plane perpendicular to the diameter through P. Shew that, if the radius of the section by the plane subtends an angle α at P, the attraction at P of the portion of the sphere which contains P is $3\gamma M \cos^2 \alpha\,(1 - \frac{2}{3}\cos\alpha)/a^2$.

[London Univ. 1937]

9. A solid right circular cylinder is of density ρ and length l, and the radius of each end is a. Prove that the potential, due to it, at the centre of each end is

$$\pi\gamma\rho\{l\,\sqrt{(a^2 + l^2)} - l^2 + a^2\sinh^{-1}(l/a)\},$$

where γ is the constant of gravitation.

If in the above cylinder $l = 8a/3$, shew that the potential at the centre of the cylinder is

$$2\pi\gamma\rho a^2\,(\tfrac{4}{9} + \log_e 3).\qquad \text{[London Univ. 1935]}$$

10. Prove that the potential of a uniform circular cylinder, of length $2a$, radius of base a, and mass M at the middle point of its axis, is

$$\gamma M\,\{\sqrt{2} - 1 + \log_e(\sqrt{2} + 1)\}/a.$$

[London Univ. 1938]

11. P is a point outside a thin uniform spherical shell at a distance c from the centre. Prove that the sphere of centre P and radius c cuts the shell into two portions whose potentials at P are equal.

[London Univ. 1933]

12. A thin uniform spherical shell of density ρ has an internal radius a and external radius $a + \alpha$, where $(\alpha/a)^2$ is negligible. Find the attraction of the shell at a distance $a + k\alpha$ from the centre, where $0 < k < 1$. Hence shew that, if the outer radius contracts so that the shell becomes a thin layer of surface density σ and radius a, the attraction of this layer at a point in itself is $4\pi\sigma k$. How does this illustrate the discontinuity of normal force in crossing a surface? [C. 1928]

13. Prove that if a solid sphere of uniform density ρ is divided into two parts by a plane which cuts the surface in a circle of radius b, the attraction between the two parts is $\frac{1}{8}\pi^2\gamma\rho^2 b^4$. [London Univ. 1936]

14. A solid homogeneous sphere, of radius R and mass M, is divided into two portions by a plane at a perpendicular distance $\frac{1}{2}R$ from the centre. Shew that the mutual attraction between the two portions of the sphere is $\gamma\,\dfrac{27}{256}\dfrac{M^2}{R^2}$. [London Univ. 1933]

15. From an infinite plate of thickness $2a$ of uniform homogeneous matter a spherical portion touching the two plane faces is cut away. Prove that the attraction of the plate at a point where the sphere touches either of its faces is $\frac{2}{3}$ of its former value.

[London Univ. 1927]

16. A uniform solid sphere of mass M is placed near an infinite plate whose surface density is uniform and equal to σ. Prove that the sphere attracts the plate with a force $2\pi\gamma M\sigma$. [London Univ. 1932]

17. Shew that, if M be the mass of a sphere of radius a, the loss of gravitational potential energy in assembling the particles from a state of diffusion at an infinite distance is $\frac{3}{5}\gamma M^2/a$. [C. 1926]

18. If the radius of a solid gravitating sphere be a, the density ρ, and an internal point P be at a distance b from the centre O, prove that the difference of the potentials at P due to the two portions into which the sphere is divided by a plane through P perpendicular to OP is

$$\frac{4\pi\rho}{3b}\{a^3 - (a^2 - b^2)^{\frac{3}{2}}\}.$$ [C. 1908]

19. The density at distance r from the centre, in a sphere of attracting matter, is $\rho_0 - \lambda r$, where ρ_0 is the density at the centre, and λ is constant. Find the potential at any internal point, and prove that the potential at the centre is

$$4\pi\gamma a^2\left(\frac{\rho_0}{6} + \frac{\rho_1}{3}\right),$$

where a is the radius of the sphere, ρ_1 the density at the surface of the sphere, and γ the constant of gravitation. [London Univ.]

20. If AB is a diameter of a spherical shell of uniform surface density, find the position of a plane perpendicular to AB which divides the shell into two portions such that their potentials at A are equal.

If the surface density of the shell is σ, find the attraction at its centre due to each portion of the shell. [London Univ. 1938]

21. Shew that the initial rate of decrease of g in descending a mine shaft would be equal to g/a if the density (ρ) of the earth (radius a) were uniform. But if the earth had a spherical nucleus of different density and of radius b, the density of this nucleus must be

$$\rho\left\{1+\frac{1-\lambda}{2+\lambda}\frac{a^3}{b^3}\right\},$$

where $\lambda g/a$ is the initial rate of decrease of g in descending the shaft. [C. 1930]

22. A sphere of radius a, mass M, and density varying directly as the distance from the centre, is built up of matter brought in small amounts from a state of complete dispersion at infinity. Shew that the work W made available during the process is given by $W = \frac{4}{7}\gamma M^2/a$, where γ is the constant of gravitation.

Shew that, if the matter is now redistributed so as to form a sphere of the same radius but of uniform density, a further amount of work $\frac{1}{20}W$ is made available. [P. 1936]

23. Find the total work done by the gravitational forces in collecting from infinite dispersion six equal particles of mass m and placing them at the corners of a regular hexagon of side $2a$. [London Univ. 1932]

24. Six particles, each of mass m, are placed symmetrically on the circumference of a circle of radius r and are allowed to move from rest under their mutual attractions. Find the velocity of each particle when they are on a circle of radius $\frac{1}{2}r$. [London Univ. 1933]

---

25. Assuming the earth to be a homogeneous sphere of density σ and radius a, surrounded by a spherical shell of density ρ and thickness $b-a$, shew that the weight of a body given by a spring balance is unaltered at first on descending vertically from the surface if $\rho = 2a^3\sigma/(2a^3+b^3)$. The rotation of the earth may be neglected. [M. T. 1914]

26. A solid sphere of radius a and uniform density is surrounded by a concentric shell of internal radius a and external radius b in which the density is a function of the distance from the centre. Find the law of density if the force of attraction is constant in the substance of the shell, and prove that the mass of the shell is to the mass of the sphere as b^2-a^2 is to a^2. [C. 1932]

27. The density of a sphere of radius a and mean density ρ is a function of the distance from the centre such that the attraction at any point inside the sphere is proportional to the square of the distance from the centre. Express the density at a distance r from the centre in terms of ρ, a and r, and prove that the potential at an internal point at a distance r from the centre is $\dfrac{4}{9}\dfrac{\pi\gamma\rho}{a}(4a^3 - r^3)$. [P. 1933]

28. If O be the centre of a solid uniform sphere and P an external point, shew that the sphere on OP as diameter cuts off a portion whose attraction at P is equal to $\dfrac{1}{2}\dfrac{M}{OP^2}$, where M is the whole mass.

[C. 1905]

29. Find the potential at any point in the interior of a sphere of radius a, the density of the sphere at a distance r from the centre being $\rho_0(1 - kr^n)$.

Shew that, if the sphere be divided into two parts by a plane through the centre, the mutual pressure between the parts is

$$\tfrac{1}{3}\pi^2\gamma\rho_0^2a^4\left[1 - \frac{4(n+6)}{(n+3)(n+4)}ka^n + \frac{6}{(n+2)(n+3)}k^2a^{2n}\right].$$

[London Univ. 1934]

30. A portion of a uniform solid paraboloid of revolution is cut off by a plane at right angles to the axis OX. Prove that the attraction at the focus S will be zero when $\sin(\tfrac{1}{2}PSX) = 1/\sqrt{e}$, P being a point on the rim. [M. T. 1905]

31. If ρ be the density of a homogeneous solid sphere and a the radius, prove that the force in astronomical units exerted by the rest of the sphere on a conical portion of itself bounded by a right circular cone of angle 2α with its vertex at the centre is $\tfrac{1}{3}\pi^2\rho^2a^4\sin^2\alpha$.

[C. 1915]

32. A solid of uniform density ρ is in the form of the spheroid obtained by rotating an ellipse, of latus rectum $2l$ and eccentricity e, about the major axis. Prove that the potential and the intensity of the attraction at a focus of the generating ellipse are

$$2\pi\gamma\rho l^2/(1 - e^2) \quad \text{and} \quad (2\pi\gamma\rho l/e^2)[\log\{(1+e)/(1-e)\} - 2e]$$

respectively. [London Univ. 1936]

33. Find the potential and attraction, at any point, of a homogeneous sphere with a spherical excentric cavity, and deduce the corresponding values for a thin spherical shell whose small thickness at any point is proportional to its distance from a plane. [C. 1899]

34. Two uniform solid spheres are of density ρ and radius a, and the distance between their centres is $c\,(>2a)$. Prove that they attract each other with a force $16\pi^2\gamma\rho^2a^6/9c^2$,

where γ is the constant of gravitation. [London Univ. 1926]

35. A solid sphere of radius a and mass M is of uniform density. Prove that the attraction on an octant of the sphere due to the rest of the sphere has components parallel to the edges of the octant each equal to $\frac{3}{64}\gamma M^2/a^2$, where γ is the gravitational constant.   [C. 1914]

36. Prove that the work done by the gravitational forces when a uniform spherical shell, of mass M and of external and internal radii a and b respectively, is changed to a solid sphere of the same density is

$$\frac{3\gamma M^2}{10c^6}[2c^5 - 2a^5 - 3b^5 + 5a^2b^3],$$

where $c^3 = a^3 - b^3$.   [London Univ. 1939]

37. The intensity of gravitational attraction at any point on the surface of a certain uniform solid sphere, of radius a and centre O, is g. A spherical cavity of radius b is made with its centre at a point C, where $OC = c$, $(b + c < a)$. Prove that the magnitudes of the intensities at the points where the line OC (produced) meets the surface of the sphere are

$$g\left\{1 - \frac{b^3}{a\,(a \pm c)^2}\right\},$$

and that at the ends of any diameter through O perpendicular to OC they are

$$g\left\{1 - \frac{2b^3}{(a^2 + c^2)^{\frac{3}{2}}} + \frac{b^6}{a^2\,(a^2 + c^2)^2}\right\}^{\frac{1}{2}}.$$

[London Univ. 1935]

38. A uniform spherical shell of gravitating matter, in which the volume of the matter is equal to that of the hollow it encloses, contracts under its own gravitation into a uniform solid sphere of the same density. Prove that the work done by the gravitating forces is $3(2^{\frac{1}{3}} - 1)\gamma M^2/10a$, where M is the mass and a the radius of the solid sphere.   [London Univ.]

39. The density of a sphere, of mass M and radius a, at a distance r from the centre is proportional to $a - r$. Prove that the potential at this distance is $\gamma M(2a^3 - 2ar^2 + r^3)/a^4$, and that the potential energy exhausted in collecting the elements of the sphere is $26\gamma M^2/35a$, where γ is the gravitational constant.   [London Univ. 1926]

40. Find the negative potential energy W for a spherical distribution of mass M and radius a in which the density varies uniformly from ρ_1 at the surface to ρ_0 at the centre, and deduce that, if $\rho_1 = 0$,

$$W = 26GM^2/35a;$$

while if $\rho_0 = 0$, $\rho_1 \neq 0$ this value of W is reduced in the ratio $10/13$. G denotes the gravitation constant.   [M. T. 1928]

41. Prove that, if a segment of a uniform thin spherical shell has an angular radius α, its attractions on the two sides, at points close to the pole of the segment, are $2\pi\gamma\sigma(1 \pm \sin\frac{1}{2}\alpha)$, where σ is the surface density.   [M. T. 1922]

42. Defining the *principal* value of the attraction at a point of an attracting surface as the limit of the attraction at the centre of a circular hole of which the radius tends to zero, shew that its tangential components are continuous with those of the attraction near the surface, while its normal component is the mean of the limiting normal components on either side.

Shew that the attraction at the focus of a small elliptical hole of eccentricity e in a thin uniform spherical shell makes an angle

$$\tan^{-1}\left(\frac{1 - \sqrt{1-e^2}}{e}\right)$$

with the radius of the shell. [C. 1929]

43. Prove that the work done by a gravitating system in attracting a new body from infinity is equal to $\frac{1}{2}\int \rho\,(V + V')\,d\tau$, where the integral is taken through the space occupied by the new body, and V, V' are the gravitational potentials before and after the new body arrives.

Calculate the work done in filling up a spherical hollow of radius b in a uniform sphere of radius a, the distance between the centres being c and the density ρ. [C. 1914]

44. Find an expression for the mutual gravitational potential energy of a uniform sphere of mass M and a uniform rod of length $2a$ and mass m when they are placed in any relative position.

Shew that this is constant when the centre of the sphere moves on the spheroid formed by rotating an ellipse whose foci are the ends of the rod about the rod. [M. T. 1923]

45. A uniform rod BC of mass M is free to turn about its middle point O which is fixed, and a uniform sphere of mass m has its centre at A. Shew that the couple on the rod tending to turn it into the line OA is $\frac{1}{2}\gamma Mm\left(\frac{|b-c|}{bc}\tan\frac{1}{2}\alpha\right)$, where b, c denote the lengths of AC, AB and α the angle BAC. [M. T. 1930]

46. The profile of a long straight mountain range of uniform density ρ is an arc of a circle less than a semicircle. Prove that if h is the height of the range and b its breadth at its base, a plumb line at its foot will be deflected towards the range through an angle approximately equal to

$$\frac{3\rho b}{2\pi\rho_0 R}\tan^{-1}\frac{2h}{b},$$

where R and ρ_0 are the radius and mean density of the earth.

[London Univ. 1927]

47. Shew that, if the potential of a particle of mass m at distance r is $m\phi'''(r)/r$, then the potential of a uniform solid sphere of mass M and radius a at an external point at distance r from the centre is

$$\frac{3M}{2a^3 r}\{a\left[\phi'(r+a) + \phi'(r-a)\right] - \left[\phi(r+a) - \phi(r-a)\right]\},$$

where the functions $\phi'(x)$, $\phi''(x)$, $\phi'''(x)$ are successive derivatives of $\phi(x)$.

Consider in particular $\phi'''(x) = \gamma e^{-\lambda x}$, where $\lambda \geqslant 0$. Shew that the external field of the sphere is the same as that of a single particle at its centre; and that the mass of this particle is

$$3M \, (\lambda a \cosh \lambda a - \sinh \lambda a)/\lambda^3 a^3$$

if $\lambda > 0$, or M if $\lambda = 0$.   [M. T. 1924]

48. Shew that the potential of a uniform circular disc at a point of itself is

$$\int_0^{2\pi} \frac{\sigma a \, (a - r \cos \theta) \, d\theta}{\sqrt{(a^2 - 2ar \cos \theta + r^2)}},$$

where r is the distance of the point from the centre and σ is the mass per unit area.

Deduce that the work done in removing all the particles to infinity is $\frac{8}{3}\pi\sigma^2 a^3$.   [C. 1924]

49. If the attraction between two elements of matter of mass m and m' is $mm' f(R)$, where R is the distance between them, shew that the inward attraction due to a uniform thin spherical shell of mass M and internal radius a, at an internal point distant $x \ (<a)$ from the centre, is F, where

$$F = \frac{M}{4ax^2} \int_{a-x}^{a+x} (R^2 + x^2 - a^2) f(R) \, dR.$$

If F is zero for all values of a and x, subject to the condition $x < a$, shew that the only possible form of $f(R)$ is A/R^2, where A is some constant.   [M. T. 1937]

Chapter IV

THEOREMS OF LAPLACE, POISSON AND GAUSS. GENERAL THEORY

4·1. Laplace's equation for the potential. (i) Let V be the potential of a system of attracting particles at a point (x, y, z) not in contact with the particles, so that $V = \Sigma m/r$, where m is the mass of the particle at (a, b, c) and

$$r^2 = (x-a)^2 + (y-b)^2 + (z-c)^2.$$

Then
$$\frac{\partial V}{\partial x} = -\Sigma \frac{m}{r^2} \frac{\partial r}{\partial x} = -\Sigma \frac{m(x-a)}{r^3},$$

and
$$\frac{\partial^2 V}{\partial x^2} = -\Sigma \frac{m}{r^3} + 3\Sigma \frac{m(x-a)^2}{r^5}.$$

Similarly
$$\frac{\partial^2 V}{\partial y^2} = -\Sigma \frac{m}{r^3} + 3\Sigma \frac{m(y-b)^2}{r^5},$$

and
$$\frac{\partial^2 V}{\partial z^2} = -\Sigma \frac{m}{r^3} + 3\Sigma \frac{m(z-c)^2}{r^5}.$$

Whence, by addition,

$$\frac{\partial^2 V}{\partial x^2} + \frac{\partial^2 V}{\partial y^2} + \frac{\partial^2 V}{\partial z^2} = 0.$$

(ii) Let V be the potential of a continuous body or bodies at a point (x, y, z) outside the body or bodies, so that $V = \int \frac{\rho \, dv}{r}$, where ρ is the density of the element of volume dv at (x', y', z'), and

$$r^2 = (x-x')^2 + (y-y')^2 + (z-z')^2.$$

Since r is measured from an external point P there is nothing to make the integral other than convergent, and the results of differentiation with regard to x, y and z under the integral sign are continuous and yield convergent integrals, therefore we may differentiate under the integral sign and write

$$\frac{\partial V}{\partial x} = -\int \frac{\rho\,(x-x')}{r^3}\,dv$$

and

$$\frac{\partial^2 V}{\partial x^2} = -\int \left\{ \frac{\rho}{r^3} - \frac{3\rho\,(x-x')^2}{r^5} \right\} dv.$$

Similarly

$$\frac{\partial^2 V}{\partial y^2} = -\int \left\{ \frac{\rho}{r^3} - \frac{3\rho\,(y-y')^2}{r^5} \right\} dv$$

and

$$\frac{\partial^2 V}{\partial z^2} = -\int \left\{ \frac{\rho}{r^3} - \frac{3\rho\,(z-z')^2}{r^5} \right\} dv.$$

Whence, as before, by addition,

$$\left. \begin{aligned} \frac{\partial^2 V}{\partial x^2} + \frac{\partial^2 V}{\partial y^2} + \frac{\partial^2 V}{\partial z^2} &= 0 \\ \nabla^2 V &= 0 \end{aligned} \right\} \quad \dots\dots\dots\dots(1).$$

or

This result is known as **Laplace's equation.*** It is satisfied by the potential of an attracting system at every point at which there is no matter.

4·11. Poisson's equation for the potential. Now let the point P of co-ordinates x, y, z be inside the attracting matter. Describe a sphere of small radius ϵ and centre (a, b, c) containing the point P, taking ϵ so small that we may regard the density ρ of the matter in this sphere as uniformly distributed. The matter which produces the potential V at P may now be divided into two parts, viz. the matter outside and the matter inside the small sphere. Let V_1, V_2 denote their contributions to the whole potential V at P. Since P is not in contact with the matter which produces the potential V_1, therefore, by **4·1**, $\nabla^2 V_1 = 0$. And V_2 being the potential at

---

* Pierre Simon, Marquis de Laplace (1749–1827), distinguished French mathematician. His great work *Mécanique Céleste* in five volumes was published between 1799 and 1825. The theorem above is to be found in *Mécanique Céleste*, t. i, p. 137, Paris, An VII.

a point (x, y, z) inside a small sphere of radius ϵ, we have, from 3·31,

$$V_2 = \tfrac{2}{3}\pi\rho\,(3\epsilon^2 - r^2),$$

where r is the distance of (x, y, z) from the centre (a, b, c),

$$= \tfrac{2}{3}\pi\rho\,\{3\epsilon^2 - (x-a)^2 - (y-b)^2 - (z-c)^2\}.$$

Hence $\dfrac{\partial V_2}{\partial x} = -\tfrac{4}{3}\pi\rho\,(x-a),$ and $\dfrac{\partial^2 V_2}{\partial x^2} = -\tfrac{4}{3}\pi\rho.$

Similarly $\dfrac{\partial^2 V_2}{\partial y^2} = \dfrac{\partial^2 V_2}{\partial z^2} = -\tfrac{4}{3}\pi\rho,$

and $\dfrac{\partial^2 V_2}{\partial x^2} + \dfrac{\partial^2 V_2}{\partial y^2} + \dfrac{\partial^2 V_2}{\partial z^2} = -4\pi\rho.$

But $V = V_1 + V_2$; therefore at every point at which there is attracting matter of volume density ρ

$$\left.\begin{aligned} \frac{\partial^2 V}{\partial x^2} + \frac{\partial^2 V}{\partial y^2} + \frac{\partial^2 V}{\partial z^2} &= -4\pi\rho \\ \nabla^2 V &= -4\pi\rho \end{aligned}\right\} \quad \dots\dots\dots\dots(1).$$

or

This result is known as **Poisson's equation.**[*]

4·12. Attraction = grad V, inside the body. We may adopt the line of argument of **4·11** to shew that the relations $X, Y, Z = \dfrac{\partial V}{\partial x}, \dfrac{\partial V}{\partial y}, \dfrac{\partial V}{\partial z}$, which we have seen to be true at points outside a system of attracting particles, are also true at all points outside or inside a continuous distribution of matter.

The definitions of V and X for a continuous distribution are

$$V = \int \frac{\rho\,dv}{r} \quad \text{and} \quad X = \int \frac{\rho\,(x'-x)}{r^3}\,dv,$$

[*] Siméon Denis Poisson (1781–1840). French mathematician and physicist, author of more than 300 papers. The proof given in **4·11** is very nearly in the form given by Poisson in *Nouveau Bulletin des Sciences par la Société Philomathique de Paris*, t. iii, p. 388; 5ᵉ Année, 1812. He observes that, in the case of a body of continuously varying density, since $\nabla^2 V_1 = 0$, it follows that $\nabla^2 V$ and also $\nabla^2 V_2$ must be independent of the form and dimensions of the element of which V_2 is the potential, so that this element may be taken to be so small that its variation in density may be neglected. An alternative proof is given in **4·21** and **4·2**.

where the symbols are as defined in **4·1**, and where for an internal point (x, y, z) the definitions need the amplification of **3·1**.

When the point (x, y, z) is outside the matter, differentiation under the integral sign is permissible, and since

$$r^2 = (x - x')^2 + (y - y')^2 + (z - z')^2,$$

$$\frac{\partial V}{\partial x} = \int \rho \frac{\partial}{\partial x} \left(\frac{1}{r} \right) dv = \int \frac{\rho(x' - x)}{r^3} dv = X.$$

But when the point (x, y, z) is inside the matter we proceed as in **4·11** and use suffixes 1 and 2 to denote contributions arising from the matter outside and the matter inside the small sphere. Since the point (x, y, z) is not in contact with the matter outside the small sphere, we have as above $X_1 = \partial V_1 / \partial x$, and for the matter inside the sphere

$$V_2 = \tfrac{2}{3}\pi\rho \left\{ 3\epsilon^2 - (x - a)^2 - (y - b)^2 - (z - c)^2 \right\}$$

and

$$X_2 = -\tfrac{4}{3}\pi\rho r \cdot \frac{x - a}{r} = -\tfrac{4}{3}\pi\rho(x - a) = \frac{\partial V_2}{\partial x}.$$

Hence, for the whole attraction component,

$$X = X_1 + X_2 = \frac{\partial V_1}{\partial x} + \frac{\partial V_2}{\partial x} = \frac{\partial V}{\partial x},$$

at points inside the attracting matter.

4·13. To summarize our results up to this point:

(i) The attraction is the gradient of a potential function V both outside and inside the attracting matter (**2·22, 4·12**).

(ii) In empty space the potential satisfies Laplace's equation $\nabla^2 V = 0$ (**4·1**).

(iii) At any point at which there is matter of volume density ρ the potential satisfies Poisson's equation $\nabla^2 V = -4\pi\rho$ (**4·11**).

(iv) When there is a surface distribution of matter the potential function assumes different forms V_1, V_2 on opposite sides of the surface, but *at* the surface they satisfy the conditions

$$V_1 = V_2$$

and
$$\frac{\partial V_2}{\partial n} - \frac{\partial V_1}{\partial n} = -4\pi m,$$

where m is the surface density of the matter, and ∂n is an element of the normal directed from region 1 to region 2 (3·7).

4·2. Gauss's theorem.* *The outward flux of the force of attraction over any closed surface in a gravitational field of force is equal to* -4π *times the mass enclosed by the surface.*

This theorem can be deduced in two steps, from Green's theorem **1·3** and Poisson's equation **4·11**.

Thus, in the notation of **1·3**, the outward normal component of force across an element dS of the closed surface is $lX + mY + nZ$ and, using dv for an element of the volume enclosed,

$$\int (lX + mY + nZ)\,dS = \int \left(\frac{\partial X}{\partial x} + \frac{\partial Y}{\partial y} + \frac{\partial Z}{\partial z}\right) dv$$

$$= \int \left(\frac{\partial^2 V}{\partial x^2} + \frac{\partial^2 V}{\partial y^2} + \frac{\partial^2 V}{\partial z^2}\right) dv$$

$$= -4\pi \int \rho\,dv \quad \text{(Poisson's equation)}$$

$$= -4\pi \quad \text{(mass enclosed).}$$

Conversely, Poisson's equation may be deduced from Gauss's theorem. For if we write $\int \rho\,dv$ for the mass enclosed by S, we have

$$-4\pi \int \rho\,dv = \int (lX + mY + nZ)\,dS \quad \text{(Gauss)}$$

$$= \int \left(\frac{\partial X}{\partial x} + \frac{\partial Y}{\partial y} + \frac{\partial Z}{\partial z}\right) dv \quad \text{(Green)}$$

$$= \int \left(\frac{\partial^2 V}{\partial x^2} + \frac{\partial^2 V}{\partial y^2} + \frac{\partial^2 V}{\partial z^2}\right) dv.$$

Therefore
$$\int \left(\frac{\partial^2 V}{\partial x^2} + \frac{\partial^2 V}{\partial y^2} + \frac{\partial^2 V}{\partial z^2} + 4\pi\rho\right) dv = 0.$$

* Carl Friedrich Gauss (1777–1855). German mathematician and physicist.

And this is true for integration through the volume enclosed by every closed surface, i.e. for all ranges of integration, so that the integrand must vanish, or $\nabla^2 V = -4\pi\rho$.*

4·21. Independent proof of Gauss's theorem. Regarding the field as due to a system of particles, let m be the mass of a particle situated at O. Let a cone of small solid angle $d\omega$ and vertex O cut a closed

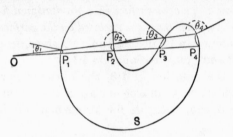

surface S in elements of area dS_1, dS_2, dS_3, ... at P_1, P_2, P_3, ..., and let the outward drawn normals at P_1, P_2, P_3, ... make angles θ_1, θ_2, θ_3, ... with the line ... $P_3 P_2 P_1 O$. Then so far as the outward flux of force from S depends on the particle m at O, the contributions of the elements dS_1, dS_2, dS_3, ... are each of the form $\dfrac{m\,dS}{PO^2}\cos\theta$. But

---

* In vector symbolism, we have the resultant attraction $\mathbf{R} = \mathbf{grad}\,V$, and the flux of $\mathbf{R}$ out of a region bounded by a closed surface S

$$= \int R_n dS$$

$$= \int \operatorname{div} \mathbf{R}\, dv \quad (1·7)$$

$$= \int \operatorname{div} \mathbf{grad}\, V dv$$

$$= \int \nabla^2 V\, dv = -4\pi \int \rho\, dv$$

$$= -4\pi \quad \text{(mass enclosed);}$$

and conversely, if

$$-4\pi \int \rho\, dv = \int R_n dS$$

$$= \int \operatorname{div} \mathbf{R}\, dv = \int \operatorname{div} \mathbf{grad}\, V dv$$

$$= \int \nabla^2 V dv,$$

then $\int (\nabla^2 V + 4\pi\rho)\, dv = 0$, for all ranges of integration, and $\nabla^2 V = -4\pi\rho$.

this expression is equal to $\pm m\,d\omega$; $+$ or $-$ according as θ is acute or obtuse, i.e. $+$ or $-$ according as the cone proceeding from O is entering or leaving the region bounded by S. It follows that when O is outside the region (as in the figure), since the cone leaves the region as often as it enters it, therefore the total contribution to the outward normal flux arising from the intersections of this cone and the surface is zero. But when O lies inside S the cone leaves the region and if it re-enters, it finally emerges, so that it leaves the region once more often than it enters it, and the total contribution in this case is $-m\,d\omega$. Hence by taking cones in all directions round O, we see that a particle of mass m contributes zero to the outward normal flux when it lies outside the surface, and $-4\pi m$ when it lies within the surface. And by summing for all particles of the system we obtain the required result, viz. total outward normal flux $= -4\pi$ (mass enclosed).

4·22. When the field is due to the attraction of a continuous body and the surface S intersects the body,

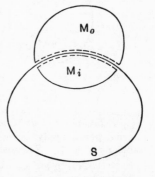

we may regard the body as divided into distinct parts of masses M_i and M_o, inside and outside S, and suppose that these are separated by an infinitesimal distance so that M_i lies wholly within S and M_o wholly without it. The outward normal flux of force from S is then $-4\pi M_i$ as before, and this will continue to be true as the distance between M_i and M_o decreases without limit.

The result is not affected by the 'matter on the surface S' because we are considering a body of finite volume density and, as explained in **3·7**, it requires a finite *surface* density of matter (i.e. an infinite volume density) to affect the normal component of force across the surface.

There is, however, an exceptional case in which the theorem needs modification, namely when the matter is a distribution of finite surface density on the surface S. Reverting to **4·21** we observe that if the point O lies on the surface S, then only cones on one side of the tangent plane at S intersect the surface, so that the contribution of the mass m to the total outward flux is $-2\pi m$, and the same being true for all matter lying on the surface, the total outward flux due to a total mass M is $-2\pi M$.

It is easy to verify, for a uniform distribution of mass M on the surface of a sphere S, that the total outward flux of force

across a concentric sphere inside S is zero (**3·2**),

across S is $-2\pi M$ (**3·71**),

and   across a concentric sphere outside S is $-4\pi M$ (**3·2**).

4·23. Applications of Gauss's Theorem to spheres and cylinders.

(a) *Spherical shell.* Let M be the mass of a uniform spherical shell of radius a. Apply Gauss's theorem to a concentric sphere S of radius r. By symmetry the outward normal force R has the same value at all points of S.

When $r < a$, fig. (i), S encloses no matter, so that

$$4\pi r^2 R = 0, \quad \text{and} \quad R = 0 \text{ at all points inside the shell.}$$

When $r > a$, fig. (ii), S encloses a mass M, so that

$$4\pi r^2 R = -4\pi M, \quad \text{and} \quad R = -M/r^2 \text{ at all points outside the shell (3·2).}$$

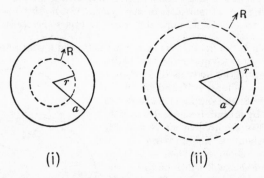

(i)   (ii)

(b) *Solid sphere of uniform density* ρ. To find the attraction at an internal point, apply Gauss's theorem to a concentric sphere S of radius $r < a$, fig. (i). S now encloses a mass $\frac{4}{3}\pi\rho r^3$, so that

$$4\pi r^2 R = -4\pi . \tfrac{4}{3}\pi\rho r^3 \quad \text{and} \quad R = -\tfrac{4}{3}\pi\rho r \text{ (3·3).}$$

The attraction at an external point may be found as for a uniform spherical shell.

(c) *Long uniform cylinder.* Let the figures now represent cross-sections of a uniform long cylinder of radius a and mass M per unit length. At points sufficiently distant from the ends of the cylinder we may neglect force parallel to the axis, and as above let R denote radial force. Then we may take for S a unit length of a coaxial cylinder of radius r and ignore any force across its plane ends.

When $r < a$, fig. (i), S encloses no matter, so that

$$2\pi r R = 0, \quad \text{and} \quad R = 0 \text{ at all points inside the cylinder.}$$

When $r > a$, fig. (ii), S encloses a mass M, so that

$$2\pi r R = -4\pi M, \text{ and } R = -2M/r \text{ at all points outside the cylinder (2·42).}$$

4·24. Tubes of force. If the lines of force (2·26) be drawn through every point of a small closed curve drawn on an

equipotential surface, these lines form a tube of which the
equipotential surface is a cross-section, called **a tube of
force**.

Apply Gauss's theorem to the region bounded by a portion
of a tube of force cut off be-
tween two equipotential surfaces and not enclosing any
matter. If R_1, R_2 denote the
force of the field at the ends of
the tube and ω_1, ω_2 the areas
of the ends, since there is no
flux of force through the sides
of the tube and no matter
inside it, therefore

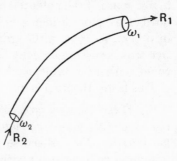

$$R_1\omega_1 - R_2\omega_2 = 0,$$

or the force varies inversely as the area of the cross-section
of the tube.

Example. If the field of force be due to the attraction of a uniform
infinite plane sheet of matter of mass m
per unit area, it is evident that the equipotential surfaces are planes parallel to
the given plane and the tubes of force are
straight, so that the force is constant. If
R denotes the force directed away from
the plane and we apply Gauss's theorem
to a portion of a tube of force of cross-
section ω extending to both sides of the
plane, since by symmetry the force has
the same value on both sides of the
plane

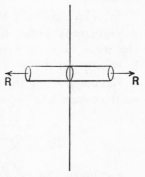

$$2R\omega = -4\pi\omega m, \quad \text{or} \quad R = -2\pi m \ (2\cdot5).$$

4·3. Theorems on the potential. (i) *The potential cannot
have a maximum or a minimum value at any point of space
unoccupied by matter.*

For if the potential had a maximum value at a point O
we could surround O by a small sphere S such that at every
point of S the potential would be less than at O. The outward
normal force R_n would therefore be negative at all points of

S, and $\int R_n \, dS$ taken over S could not vanish, as it must if S contains no attracting matter. Similarly for a minimum.

As a consequence of this theorem we note that at a point in free space, if the potential be not constant in the neighbourhood of the point it increases in some directions and decreases in others, so that, although a particle placed at the point free from constraint might be in equilibrium its equilibrium could neither be stable, nor unstable for all displacements.

This latter theorem is due to Earnshaw.

(ii) *If the potential be constant over a closed surface containing none of the attracting matter, it must be constant throughout the interior.* For otherwise it would have a maximum or minimum value in the interior, which is impossible by (i).

(iii) *Gauss's Mean Value Theorem. The mean value over a spherical surface of the potential of any attracting system is equal to the potential at the centre of the sphere due to the mass outside the sphere, plus the mass inside the sphere divided by the radius.*

Let a be the radius of the sphere, m the mass of a particle at an external point A and m' the mass of a particle at an internal point A'.

The potential at any point Q on the sphere is

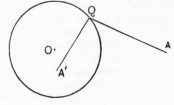

$$V = \frac{m}{AQ} + \frac{m'}{A'Q}.$$

The mean value of V over the sphere is $\dfrac{1}{4\pi a^2} \int V \, dS$, where dS denotes an element of area at Q.

This $\qquad = \dfrac{m}{4\pi a^2} \int \dfrac{dS}{AQ} + \dfrac{m'}{4\pi a^2} \int \dfrac{dS}{A'Q}.$

But $\int \dfrac{dS}{AQ}$ is the potential at A due to a spherical layer of unit surface density and therefore $= 4\pi a^2/OA$.

Similarly, $\int \dfrac{dS}{A'Q}$ is the potential at A' due to the same layer and therefore $= 4\pi a^2/a = 4\pi a$.

Hence the mean value of $V = \dfrac{m}{OA} + \dfrac{m'}{a}$.

In like manner the mean value of the potential due to any attracting system $= \Sigma \dfrac{m}{OA} + \dfrac{\Sigma m'}{a}$, where the first sum represents the potential at O due to the matter outside the sphere, and thus the theorem is proved.

(iv) *If the potential be constant throughout any region* T, *unoccupied by matter, it has the same constant value through all space that can be reached from* T *without passing through matter* (Gauss).

Let V be the value of the potential throughout T. If the potential differs from V in neighbouring space there must be some regions contiguous to T where the potential is greater than V and others where it is less. In the neighbourhood of a place where the potential is greater than V draw a sphere with its centre O inside T. Then on the part of the sphere inside T the potential is V and on the part outside it is greater than V, so that the mean value of the potential over the sphere exceeds V, the value at the centre O of the sphere, which is impossible by the last theorem since the sphere contains no matter. Hence the region T of constant potential must be extended to include the whole of the sphere, and so on step by step through all space that does not include attracting matter.

(v) *In any case of symmetry about an axis, if the potential be constant through any finite distance along the axis, however short, not within the attracting matter, it has the same constant value through all space that can be reached from this portion of the axis without passing through matter.** Without loss of generality we may suppose that the potential is zero at all points of a segment of the axis. Let P be a point of the

---

* Thomson and Tait, *Treatise on Natural Philosophy*, Art **498**, Oxford 1867. The proof given above is due to Dr S. Verblunsky.

segment. On account of the axial symmetry, there is no force in any direction through P perpendicular to the axis. Hence the axis is normal at P to a surface on which V is constant; i.e. $V = V_P = 0$. Since the argument applies to any point P of the segment, it follows that $V = 0$ throughout a region. The theorem therefore follows from (iv).

4·31. Comparison theorems. (i) *If two different distributions of matter have equal potentials over any closed surface not including any attracting matter, they have equal potentials throughout the space enclosed and through all external space which can be reached without traversing any of the attracting matter.*

Let the attraction of one of the distributions be changed to repulsion. This changes the sign of its potential and makes the potential due to the two distributions taken together zero over the given surface. Therefore by **4·3** (ii) the potential is zero throughout the space enclosed, and by **4·3** (iv) zero throughout all external space which can be reached without passing through matter. Then changing back from repulsion to attraction, it follows that the potentials of the two distributions are equal.

(ii) *If two different distributions of matter produce equal potentials over any surface enclosing both, they produce equal potentials throughout all space external to the surface.*

Let the attraction of one of the distributions be changed to repulsion. The two distributions combined then produce zero potential over the given surface, and they also produce zero potential over the infinite sphere. Therefore their combined potential is zero throughout the space between the given surface and the infinite sphere, for this space contains none of the matter (**4·3** (ii)). Reverting to the original distributions, it follows that they have equal potentials throughout the same region.

(iii) *If two finite distributions of matter have the same equipotential surfaces outside both distributions, their attractions at every external point have the same direction and are proportional to the masses of the distributions.*

The attractions at any external point are in the same direction because they are along the inward normal to the equipotential surface through the point.

Let V, V' denote the potentials of the two distributions. Since V' is constant whenever V is constant, there must be a functional relation between them; say $V' = f(V)$. Therefore

$$\frac{\partial V'}{\partial x} = f'(V)\frac{\partial V}{\partial x}$$

and
$$\frac{\partial^2 V'}{\partial x^2} = f''(V)\left(\frac{\partial V}{\partial x}\right)^2 + f'(V)\frac{\partial^2 V}{\partial x^2}.$$

Adding the similar results obtained by differentiating with regard to y and z, we get

$$\nabla^2 V' = f''(V)\left\{\left(\frac{\partial V}{\partial x}\right)^2 + \left(\frac{\partial V}{\partial y}\right)^2 + \left(\frac{\partial V}{\partial z}\right)^2\right\} + f'(V)\nabla^2 V.$$

But, outside the matter,

$$\nabla^2 V = \nabla^2 V' = 0.$$

Therefore $\quad f''(V)\left\{\left(\frac{\partial V}{\partial x}\right)^2 + \left(\frac{\partial V}{\partial y}\right)^2 + \left(\frac{\partial V}{\partial z}\right)^2\right\} = 0.$

Hence, unless V is constant everywhere to make the second factor vanish, we have
$$f''(V) = 0,$$
giving, on integration,

$$V' = f(V) = AV + B,$$

where A, B are constants. But V and V' vanish at infinity in the ratio of the masses M, M' of the distributions, so that $B = 0$ and $A = M'/M$. Hence $V' = \dfrac{M'}{M}V$, and the result follows.

4·32. Condition that a family of surfaces is a possible family of equipotential surfaces in free space. A family of surfaces in general is not a family of equipotential surfaces unless its equation satisfies a certain condition. To find the condition that the equation

$$f(x, y, z) = \text{const.}$$

may represent a family of equipotential surfaces.

If the potential V is constant whenever $f(x, y, z)$ is constant, there must be a functional relation between V and $f(x, y, z)$; say

$$V = \phi\{f(x, y, z)\}.$$

Then
$$\frac{\partial V}{\partial x} = \phi'(f) \frac{\partial f}{\partial x}$$

and
$$\frac{\partial^2 V}{\partial x^2} = \phi''(f) \left(\frac{\partial f}{\partial x}\right)^2 + \phi'(f) \frac{\partial^2 f}{\partial x^2}.$$

By adding similar results in y and z, we get

$$\nabla^2 V = \phi''(f) \left\{\left(\frac{\partial f}{\partial x}\right)^2 + \left(\frac{\partial f}{\partial y}\right)^2 + \left(\frac{\partial f}{\partial z}\right)^2\right\} + \phi'(f) \left\{\frac{\partial^2 f}{\partial x^2} + \frac{\partial^2 f}{\partial y^2} + \frac{\partial^2 f}{\partial z^2}\right\}.$$

But in free space $\nabla^2 V = 0$; therefore

$$\frac{\dfrac{\partial^2 f}{\partial x^2} + \dfrac{\partial^2 f}{\partial y^2} + \dfrac{\partial^2 f}{\partial z^2}}{\left(\dfrac{\partial f}{\partial x}\right)^2 + \left(\dfrac{\partial f}{\partial y}\right)^2 + \left(\dfrac{\partial f}{\partial z}\right)^2} = -\frac{\phi''(f)}{\phi'(f)} = \text{a function of } f, \text{ say } \chi(f)$$
$$\dots\dots(1).$$

This is the necessary condition, and when it is satisfied the potential V can be expressed in terms of $f(x, y, z)$ thus:

We have $V = \phi(f)$, where

$$\frac{\phi''(f)}{\phi'(f)} + \chi(f) = 0.$$

Therefore
$$\log \phi'(f) = A - \int \chi(f) \, df$$

or
$$\phi'(f) = A e^{-\int \chi(f) df}.$$

It follows that
$$V = \phi(f) = A \int e^{-\int \chi(f) df} \, df + B \quad \dots\dots\dots(2).$$

4·33. Example. *Shew that a family of right circular cones with a common axis and vertex is a possible family of equipotential surfaces, and find the potential function.*

Taking the axis of z for the common axis, the equation of the family of cones is

$$f(x, y, z) \equiv \frac{x^2 + y^2}{z^2} = \text{const.} \quad \dots\dots\dots\dots(1).$$

The condition 4·32 (1) in this case becomes

$$\frac{\phi''(f)}{\phi'(f)} = -\frac{\dfrac{2}{z^2} + \dfrac{2}{z^2} + \dfrac{6\,(x^2+y^2)}{z^4}}{\dfrac{4x^2}{z^4} + \dfrac{4y^2}{z^4} + \dfrac{4\,(x^2+y^2)^2}{z^6}} = -\frac{2+3f}{2f(1+f)};$$

and the condition is satisfied since the expression on the right has reduced to a function of f. Hence

$$\frac{\phi''(f)}{\phi'(f)} + \frac{1}{f} + \frac{1}{2\,(f+1)} = 0,$$

and, by integration,

$$\log \phi'(f) + \log f + \tfrac{1}{2}\log(f+1) = \text{const.}$$

or

$$d\phi = \frac{C\,df}{f\,\sqrt{(f+1)}}.$$

Put $f = \tan^2 \theta$, and we get $d\phi = \dfrac{2C\,d\theta}{\sin \theta}.$

Hence $V \equiv \phi(f) = 2C \log \tan \tfrac{1}{2}\theta + C'.$

It is clear that θ is the half-angle of a cone, and V is constant when θ is constant.

The truth of this proposition is evident at once if we consider Laplace's equation in polar co-ordinates, namely

$$\frac{1}{r^2}\frac{\partial}{\partial r}\left(r^2\frac{\partial V}{\partial r}\right) + \frac{1}{r^2\sin \theta}\frac{\partial}{\partial \theta}\left(\sin \theta \frac{\partial V}{\partial \theta}\right) + \frac{1}{r^2\sin^2 \theta}\frac{\partial^2 V}{\partial \phi^2} = 0,$$

and whether it can have a solution which makes V constant over the surface of a right circular cone. There is clearly such a solution if V is independent of r and ϕ and satisfies

$$\sin \theta \frac{\partial V}{\partial \theta} = A,$$

where A is a constant, and this leads to

$$V = A \log \tan \tfrac{1}{2}\theta + B,$$

as before.

4·4. When the potential is given, to find the distribution of matter. We have seen that potential functions may become infinite at points or lines where matter is concentrated, and that otherwise potential functions are finite and continuous, even when crossing surfaces on which matter is concentrated (3·7); but in crossing such a surface there is a discontinuity in the gradient of the potential. Thus a closed surface on which matter is concentrated divides space into regions in which the potential function takes different forms, though these forms take equal values on the surface itself.

When the potential is given at all points of space, Poisson's equation, in the form $\rho = -\dfrac{1}{4\pi}\nabla^2 V$, serves to determine the volume density of matter wherever a finite volume density exists. If the potential is given by different functions V_1, V_2 on opposite sides of a surface S, this implies that there is on S matter of surface density σ determined by the relation

$$\sigma = -\frac{1}{4\pi}\left(\frac{\partial V_2}{\partial n} - \frac{\partial V_1}{\partial n}\right) \quad (3\cdot 7),$$

where the direction of the normal ∂n is from 1 to 2.

If the potential becomes infinite at a point, the amount of matter concentrated there can be found by applying Gauss's theorem to a small sphere having its centre at the point.

If the potential becomes infinite at all points of a line, the line density of the matter can be found by applying Gauss's theorem to a short cylinder whose axis is the line and whose radius tends to zero.

4·41. Examples. (i) *The potential outside a certain cylindrical boundary is zero; inside it is $V = x^3 - 3xy^2 - ax^2 + 3ay^2$. Find the distribution of matter.*

We have first to find the boundary. Since the potential is continuous across the boundary and zero outside, the boundary must be given by

$$V_1 = x^3 - 3xy^2 - ax^2 + 3ay^2 = 0,$$

or $(x - a)(x - \sqrt{3}y)(x + \sqrt{3}y) = 0,$

i.e. the section is an equilateral triangle OAB of height a. Then

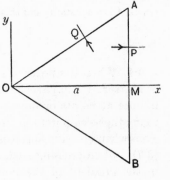

$$\frac{\partial^2 V_1}{\partial x^2} = 6x - 2a, \quad \frac{\partial^2 V_1}{\partial y^2} = -6x + 6a,$$

and $\dfrac{\partial^2 V_1}{\partial z^2} = 0,$

so that inside the prism

$$\rho = -\frac{1}{4\pi}\nabla^2 V_1 = -\frac{a}{\pi};$$

and outside $\rho = 0$ since $V_2 = 0$.

At P on AB $\sigma = -\dfrac{1}{4\pi}\left(\dfrac{\partial V_2}{\partial x} - \dfrac{\partial V_1}{\partial x}\right)_{x=a}$

$$= \dfrac{1}{4\pi}(a^2 - 3y^2) = \dfrac{3}{4\pi}(MA^2 - MP^2),$$

or $\sigma = \dfrac{3}{4\pi} AP . PB.$

At Q on AO $(x = \sqrt{3}y)$

$$\sigma = \frac{1}{4\pi}\frac{\partial V_1}{\partial n} = \frac{1}{4\pi}\left\{-\frac{1}{2}\frac{\partial V_1}{\partial x} + \frac{\sqrt{3}}{2}\frac{\partial V_1}{\partial y}\right\}_{x=\sqrt{3}y}$$

$$= \frac{1}{4\pi}\{-\tfrac{3}{2}x^2 + \tfrac{3}{2}y^2 + ax - 3\sqrt{3}xy + 3\sqrt{3}ay\}_{x=\sqrt{3}y}$$

$$= \frac{1}{\pi}x(a-x),$$

or $\sigma = \dfrac{3}{4\pi} OQ . QA$

and similarly on OB. This shews that a solid prism of uniform density a/π would produce the same external field as a distribution of matter of surface density $3AP.PB/4\pi$ on each of the faces of the prism.

(ii) *Use the theorems of Laplace and Gauss to verify that the function* $V = m\log\dfrac{r+r'+2a}{r+r'-2a}$ *is the potential due to a uniform line density m on the straight line joining the points A, B from which r, r' are measured.*

Take the origin at the middle point of AB and the axis Ox along BA, and let ζ denote distance from Ox.

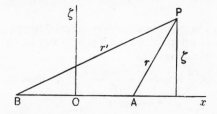

We notice that V is a finite and continuous function except on the line between B and A where it becomes infinite.

We have $r^2 = (x-a)^2 + \zeta^2, \quad r'^2 = (x+a)^2 + \zeta^2,$

and $V = m\log\left(\dfrac{r+r'+2a}{r+r'-2a} \cdot \dfrac{-r+r'+2a}{-r+r'+2a}\right)$

$$= m\log\frac{-r^2+r'^2+4a^2+4ar'}{-r^2+r'^2-4a^2+4ar} = m\log\frac{r'+x+a}{r+x-a}.$$

Hence
$$\frac{\partial V}{\partial x} = \frac{m}{r'+x+a}\left(\frac{x+a}{r'}+1\right) - \frac{m}{r+x-a}\left(\frac{x-a}{r}+1\right)$$

$$= \frac{m}{r'} - \frac{m}{r};$$

and
$$\frac{\partial^2 V}{\partial x^2} = -\frac{m(x+a)}{r'^3} + \frac{m(x-a)}{r^3} \quad \dots\dots\dots\dots(1).$$

Similarly,
$$\frac{\partial V}{\partial \zeta} = \frac{m}{r'+x+a}\cdot\frac{\zeta}{r'} - \frac{m}{r+x-a}\cdot\frac{\zeta}{r}$$

$$= \frac{m}{\zeta}\left(\frac{r'-x-a}{r'} - \frac{r-x+a}{r}\right)$$

or
$$\zeta\frac{\partial V}{\partial \zeta} = m\left(-\frac{x+a}{r'} + \frac{x-a}{r}\right) \quad \dots\dots\dots\dots(2);$$

and
$$\frac{1}{\zeta}\frac{\partial}{\partial \zeta}\left(\zeta\frac{\partial V}{\partial \zeta}\right) = \frac{m(x+a)}{r'^3} - \frac{m(x-a)}{r^3} \quad \dots\dots\dots(3).$$

By adding (1) and (3) we get
$$\frac{\partial^2 V}{\partial x^2} + \frac{1}{\zeta}\frac{\partial}{\partial \zeta}\left(\zeta\frac{\partial V}{\partial \zeta}\right) = 0,$$

shewing that, since V is symmetrical about Ox, it satisfies Laplace's equation in cylindrical co-ordinates ζ, ϕ, x. This holds good at every point save at points between A and B on the line AB where V is infinite, so there is no volume density of matter; and since there is no discontinuity in the form of V there is no surface density. Hence the matter is confined to the line AB. It remains to shew that the line density is uniform.

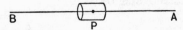

Apply Gauss's theorem to a small cylinder of length ϵ and radius ζ with its centre at any point P on AB and its axis along AB. The flux across the curved surface is

$$2\pi\epsilon\zeta\frac{\partial V}{\partial \zeta} = 2\pi\epsilon m\left(-\frac{x+a}{r'} + \frac{x-a}{r}\right), \quad \text{from (2),}$$

$$\to -4\pi\epsilon m \quad \text{as} \quad \zeta\to 0;$$

and the flux across the plane ends is $\pi\zeta^2\epsilon\dfrac{\partial^2 V}{\partial x^2}$, which tends to zero with ζ. It follows from Gauss's theorem that the mass on any element of AB of length ϵ is $m\epsilon$, so that the line density is uniform and equal to m.

4·5. Simple applications of Laplace's equation.

(i) *Field due to a uniform plane sheet.* Taking the axis of z at right angles to the sheet, V is clearly a function of z only, so that Laplace's equation reduces to $d^2V/dz^2 = 0$, which gives on integration $dV/dz = A$, where A is the constant value of the force in the direction in which z increases; and, by 2·5 or 4·24, this is $-2\pi m$, where m is the mass per unit area of the sheet, so that

$$\frac{dV}{dz} = -2\pi m$$

and $\qquad V = -2\pi mz + \text{const.} \qquad (2\cdot51).$

(ii) *Symmetry about a point.* When the field is symmetrical in such a way that V is a function of r only, in three dimensions, we can use the transformation of Laplace's equation into polar co-ordinates given in 1·68, putting $\partial V/\partial\theta = 0$ and $\partial V/\partial\phi = 0$, or we can obtain the required form directly by applying Gauss's theorem to the space between concentric spheres of radii r and $r+dr$. Assuming that this space contains no matter and that the force at distance r is dV/dr, the flux of force out of the region across the surface of the inner sphere is $-4\pi r^2\dfrac{dV}{dr}$.

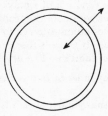

The outward flux across the sphere of radius $r+dr$ is therefore

$$4\pi r^2\frac{dV}{dr} + \frac{d}{dr}\left(4\pi r^2\frac{dV}{dr}\right)dr$$

to the first power of dr, and the vanishing of the total outward flux gives

$$\frac{d}{dr}\left(r^2\frac{dV}{dr}\right) = 0.$$

Hence, by integration, $r^2\dfrac{dV}{dr} = A$, or $\dfrac{dV}{dr} = \dfrac{A}{r^2}$,

and $\qquad V = -\dfrac{A}{r} + B \qquad \dotfill (1),$

where A and B are arbitrary constants to be determined to suit each particular problem. For example, consider the field due to a uniform thin spherical shell of radius a and mass m per unit area. Denote the potentials inside and outside the shell by V_1 and V_2. The conditions to be satisfied by V_1 and V_2 are given in 4·13, namely

where $r < a$, $\nabla^2V_1 = 0$, so that from (1), $V_1 = -\dfrac{A_1}{r} + B_1$;

where $r > a$, $\nabla^2V_2 = 0$, $\dotfill$ $V_2 = -\dfrac{A_2}{r} + B_2$;

where $r = a$, $V_1 = V_2$ and $\dfrac{dV_2}{dr} - \dfrac{dV_1}{dr} = -4\pi m$.

The last two conditions give

$$-\frac{A_1}{a} + B_1 = -\frac{A_2}{a} + B_2 \quad\text{......................}(2)$$

and

$$\frac{A_2}{a^2} - \frac{A_1}{a^2} = -4\pi m \quad\text{..........................}(3).$$

We also know that, for large values of r, V tends to zero like $4\pi ma^2/r$, so that $B_2 = 0$ and $A_2 = -4\pi ma^2$. Hence from (3) $A_1 = 0$, and from (2) $B_1 = 4\pi ma$; so that $V_1 = 4\pi ma$ and $V_2 = 4\pi ma^2/r$.

(iii) *Symmetry about an axis.* When the field is such that V is a function of r only, where r denotes distance from an axis; we may either quote the transformation of Laplace's equation into cylindrical co-ordinates given in **1·68**, putting $\partial V/\partial\theta = 0$ and $\partial V/\partial z = 0$, or we may apply Gauss's theorem to unit length of the space bounded by cylinders of radii r and $r + dr$. In this case, the flux out of this region across the cylinder of radius r is $-2\pi r\dfrac{dV}{dr}$, so that the corresponding flux across the cylinder of radius $r + dr$ is $2\pi r\dfrac{dV}{dr} + \dfrac{d}{dr}\left(2\pi r\dfrac{dV}{dr}\right)dr$ to the first power of dr, and since there is no matter in this region

$$\frac{d}{dr}\left(r\frac{dV}{dr}\right) = 0, \quad\text{or}\quad r\frac{dV}{dr} = A,$$

so that
$$V = A\log r + B,$$

where A, B are arbitrary constants to be determined to suit each particular problem.

If the matter be a long uniform thin cylinder of radius a and mass m per unit area, we may take for the internal and external potentials

$$V_1 = A_1\log r + B_1, \quad V_2 = A_2\log r + B_2.$$

When $r = a$, we have $V_1 = V_2$, so that

$$A_1\log a + B_1 = A_2\log a + B_2 \quad\text{..................}(1);$$

and when $r = a$, $\dfrac{dV_2}{dr} - \dfrac{dV_1}{dr} = -4\pi m$, so that

$$\frac{A_2}{a} - \frac{A_1}{a} = -4\pi m \quad\text{........................}(2).$$

Also applying Gauss's theorem, as in 4·23 (c), to a cylinder of radius $r\,(>a)$, we have
$$2\pi r\frac{dV_2}{dr} = -4\pi\,.\,2\pi ma,$$

so that $A_2 = -4\pi ma$.

Then from (2) $A_1 = 0$, and from (1) $B_1 = -4\pi ma\log a + B_2$, shewing that the forms for V_1 and V_2 are in agreement with the results of 2·42.

4·6. Potential of a body at a distant point. MacCullagh's Formula.* Let G be the centre of gravity of the body and P a point whose distance R from G is large compared with the dimensions of the body. Let m be the mass of an element of the body, situated at Q, where $GQ=r$ and the angle $QGP=\theta$.

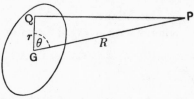

The potential at P is given by

$$V=\Sigma\frac{m}{PQ}=\Sigma\frac{m}{(R^2-2Rr\cos\theta+r^2)^{\frac{1}{2}}}$$

$$=\Sigma\frac{m}{R}\left(1-\frac{2r}{R}\cos\theta+\frac{r^2}{R^2}\right)^{-\frac{1}{2}}$$

$$=\Sigma\frac{m}{R}\left(1+\frac{r\cos\theta}{R}-\frac{r^2}{2R^2}+\frac{3r^2\cos^2\theta}{2R^2}+\ldots\right)$$

$$=\frac{\Sigma m}{R}+\frac{\Sigma mr\cos\theta}{R^2}+\frac{2\Sigma mr^2-3\Sigma mr^2\sin^2\theta}{2R^3}+\ldots.$$

But if M denotes the mass of the body, A, B, C its principal moments of inertia at G and I its moment of inertia about GP, we have

$$\Sigma m=M,\quad \Sigma mr\cos\theta=0\quad\text{(since G is the centre of gravity),}$$

$$2\Sigma mr^2=A+B+C\quad\text{and}\quad \Sigma mr^2\sin^2\theta=I.$$

Therefore
$$V=\frac{M}{R}+\frac{A+B+C-3I}{2R^3}\quad\ldots\ldots\ldots\ldots(1)$$

is an approximation to the value of the potential.

4·61. Attraction components at a distant point. Taking axes through the centre of gravity, let x,y,z be the co-ordinates of the distant point and r its distance. In **4·6** this means that

* James MacCullagh (1809–1847). Irish mathematician and physicist.

GP is now of length r and its direction cosines are x/r, y/r, z/r, so that
$$I = (Ax^2 + By^2 + Cz^2)/r^2,$$
and
$$V = \frac{M}{r} + \frac{1}{2r^5}\{(B+C-2A)\,x^2 + (C+A-2B)\,y^2$$
$$+ (A+B-2C)\,z^2\} \quad \ldots\ldots(1),$$

to this order of approximation.

The attraction components at P are therefore given by
$$X = \frac{\partial V}{\partial x} = -\frac{Mx}{r^3} + \frac{(B+C-2A)\,x}{r^5}$$
$$-\frac{5}{2}\frac{x}{r^7}\{(B+C-2A)\,x^2 + (C+A-2B)\,y^2 + (A+B-2C)\,z^2\}$$
$$\ldots\ldots(2),$$

and similar expressions for Y and Z.

Since action and reaction are equal and opposite it follows that a unit particle at P exerts on the body M forces equal and opposite to X, Y, Z acting at P. To measure their effect on the body it is convenient to transfer them parallel to themselves to act at G by introducing suitable couples. Thus the reaction on the body is represented by a force at G with components $-X$, $-Y$, $-Z$, and a couple with components

$$\left. \begin{array}{l} zY - yZ = \dfrac{3\,(C-B)\,yz}{r^5} \\[2mm] xZ - zX = \dfrac{3\,(A-C)\,zx}{r^5} \\[2mm] yX - xY = \dfrac{3\,(B-A)\,xy}{r^5} \end{array} \right\} \quad \ldots\ldots\ldots\ldots(3).$$

4·62. As an example of 4·61 (3), if the body be the earth regarded as an oblate spheroid, and a distant body be regarded as a particle of mass M' at distance r and angular distance δ above the equator, we may without loss of generality suppose the distant body to be in the plane zx, then the couple exerted by the distant body on the earth tending to cause rotation about a diameter (the y axis) is
$$\frac{3M'(A-C)\,xz}{r^5}, \quad \text{or} \quad \frac{3M'(A-C)}{r^3}\sin\delta\cos\delta,$$

and this has its greatest value when the declination δ is $\frac{1}{4}\pi$.

4·63. Mutual potential energy of two bodies at a great distance apart. Let M, M' be the masses and G, G' the centres of gravity of the bodies, and let $GG' = r$. Then if m' is the mass of an element of the second body at P and V the potential of the first

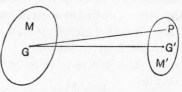

body at P, the mutual potential energy is, by **3·8**, equal to

$$-\Sigma m'V = -\Sigma m'\left\{\frac{M}{GP} + \frac{A+B+C-3I_P}{2GP^3}\right\}, \quad \text{from 4·6,}$$

where I_P denotes the moment of inertia of the body M about GP; and this

$$= -M\Sigma\frac{m'}{GP} - \frac{A+B+C-3I}{2r^3}\Sigma m',$$

correct to the order $\dfrac{1}{r^3}$, where I is the moment of inertia of the body M about GG'.

But $\Sigma\dfrac{m'}{GP}$ is the potential at G of the body M', and by **4·6** is equal to $\dfrac{M'}{r} + \dfrac{A'+B'+C'-3I'}{2r^3}$, where A', B', C', I' bear the same relation to the body M' that A, B, C, I do to the body M.

Hence we get for the mutual potential energy

$$-\left[M\left\{\frac{M'}{r} + \frac{A'+B'+C'-3I'}{2r^3}\right\} + \frac{M'(A+B+C-3I)}{2r^3}\right]$$

$$= -\left[\frac{MM'}{r} + \frac{M(A'+B'+C'-3I')}{2r^3} + \frac{M'(A+B+C-3I)}{2r^3}\right].$$

4·7. Centrobaric bodies. When the action of terrestrial or other gravity on a rigid body is reducible to a single force always acting through a point fixed relatively to the body, that point is called the **centre of gravity** of the body and the body is said to be **centrobaric**.

It follows that uniform spheres and spherical shells are centrobaric. In elementary statics it is usual to assume that the attractions of the earth on the elements of a body are a set of parallel forces with a resultant acting through a point fixed relatively to the body, so that on this hypothesis every body has a centre of gravity and its position is defined by relations

$$M\overline{x}=\Sigma mx, \quad M\overline{y}=\Sigma my, \quad M\overline{z}=\Sigma mz \quad \text{......(1)}.$$

We shall now prove the following theorem: *If the resultant attraction of a body at all external points passes through the same fixed point, that point must be the centre of gravity, as defined by* (1), *and the body must be such that every axis through the centre of gravity is a principal axis and all principal moments of inertia are equal.*

Take the fixed point as origin O. Let μ be an element of mass of the body at Q whose co-ordinates are x, y, z. Let P be an external point at distance r from O and let l, m, n be the direction cosines of OP. Then for the potential at P we have

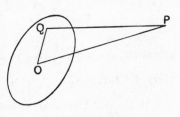

$$V=\Sigma \frac{\mu}{PQ}=\Sigma \frac{\mu}{\{(lr-x)^2+(mr-y)^2+(nr-z)^2\}^{\frac{1}{2}}}$$

$$=\Sigma \frac{\mu}{r}\left\{1-\frac{2(lx+my+nz)}{r}+\frac{x^2+y^2+z^2}{r^2}\right\}^{-\frac{1}{2}}$$

$$=\frac{M}{r}+\frac{\Sigma\mu(lx+my+nz)}{r^2}-\frac{1}{2}\frac{\Sigma\mu(x^2+y^2+z^2)}{r^3}$$

$$+\frac{3}{2}\frac{\Sigma\mu(lx+my+nz)^2}{r^3}+\dots$$

$$=\frac{M}{r}+\frac{M(l\overline{x}+m\overline{y}+n\overline{z})}{r^2}+\frac{1}{2r^3}\{l^2\Sigma\mu(2x^2-y^2-z^2)+\dots+\dots$$

$$+6mn\Sigma\mu yz+\dots+\dots\}+\dots$$

$$=\frac{M}{r}+\frac{M(l\overline{x}+m\overline{y}+n\overline{z})}{r^2}+\frac{1}{2r^3}\{(B+C-2A)l^2+\dots+\dots$$

$$+6Dmn+\dots+\dots\}+\dots,$$

where A, B, C, D, E, F are the principal moments and the products of inertia.

The resultant force at P is to pass through O for all positions of P, so that, if we write $l = \cos\theta$, $m = \sin\theta\cos\phi$, $n = \sin\theta\sin\phi$, we must have $\partial V/\partial\theta$ and $\partial V/\partial\phi$ both zero for all values of θ, ϕ. This can only be so if the coefficients of the different powers and products of l, m, n vanish separately, so that

$$\bar{x} = \bar{y} = \bar{z} = 0, \quad A = B = C \quad \text{and} \quad D = E = F = 0,$$

which proves the proposition.

4·8. Points of equilibrium. If in the presence of attracting bodies the resultant force on a free particle is zero, the position of the particle is a *point of equilibrium*. If V denotes the potential at the point P of co-ordinates x, y, z, then P is a point of equilibrium if

$$\frac{\partial V}{\partial x} = \frac{\partial V}{\partial y} = \frac{\partial V}{\partial z} = 0 \quad\ldots\ldots\ldots\ldots\ldots(1),$$

for these are the conditions that the resultant force at P is zero.

Regarding V as a function of x, y, z, the equation

$$V(x, y, z) = \text{const.} \quad\ldots\ldots\ldots\ldots\ldots(2)$$

represents a family of equipotential or level surfaces. Using suffixes to denote partial differentiation, the direction cosines of the normal at (x, y, z) on such a surface are proportional to V_x, V_y, V_z, which shews that the resultant force at the point is normal to the surface (2·26). It follows that if the point is a point of equilibrium, it is a singular point on the equipotential surface; i.e. there is not a unique tangent plane at the point, but *every* straight line through the point cuts the surface (2) in two coincident points.

Let V' be the potential at a neighbouring point P' of co-ordinates $x + \xi$, $y + \eta$, $z + \zeta$, then

$$V' = V + V_x\xi + V_y\eta + V_z\zeta + \tfrac{1}{2}(V_{xx}\xi^2 + V_{yy}\eta^2 + V_{zz}\zeta^2$$
$$+ 2V_{yz}\eta\zeta + 2V_{zx}\zeta\xi + 2V_{xy}\xi\eta) + \ldots \quad\ldots\ldots(3),$$

where the derivatives denote values at (x, y, z).

The result (3) incidentally affords another proof of the theorem **4·3** (i) that the potential cannot have a maximum or a minimum value at any point unoccupied by matter. For a first condition that V' may be greater (say) than V for all small values of ξ, η, ζ (positive or negative) is the vanishing of V_x, V_y, V_z, and the further conditions require that V_{xx}, V_{yy}, V_{zz} shall all be positive; which is not possible since $\nabla^2 V = 0$. At a point of equilibrium there will therefore be some directions in which $V' > V$ and others in which $V' < V$, so that equilibrium would be stable for some displacements and unstable for others [**4·3** (i)], and these will be separated by the surface for which $V' = V$; and from (3), since $V_x = V_y = V_z = 0$, this is, for small values of ξ, η, ζ, a quadric cone

$$V_{xx}\xi^2 + V_{yy}\eta^2 + V_{zz}\zeta^2 + 2V_{yz}\eta\zeta + 2V_{zx}\zeta\xi + 2V_{xy}\xi\eta = 0 \quad \ldots\ldots(4).$$

Further, since $V_{xx} + V_{yy} + V_{zz} = 0$, it is a cone with three perpendicular generators.

If two sheets of the same level surface intersect one another along a line, every point of this line is a point of equilibrium, because there are two normals to the surface at such a point each of which would be the direction of the resultant force if any. In such a case the tangent cone to the surface, represented by (4), breaks into two planes; so that (4) is equivalent to

$$(l\xi + m\eta + n\zeta)(l'\xi + m'\eta + n'\zeta) = 0,$$

where

$$ll' + mm' + nn' = V_{xx} + V_{yy} + V_{zz} = 0,$$

so that, if two sheets of a level surface intersect, they intersect at right angles.

4·9. Examples. (i) *A particle is placed on one of the plane faces of a uniform circular cylinder at a small distance from the centre of that face; prove that it will make small oscillations of period* $2\left(\dfrac{\pi}{\mu\rho h}\right)^{\frac{1}{2}}(a^2 + h^2)^{\frac{1}{4}}$, *where a is the radius of the cylinder, h its height and ρ its density, and μ is the attraction between two unit masses at unit distance apart.*

[M. T. 1889]

The attraction of the cylinder at an external point on its axis is found from 2·52 to be

$$2\pi\mu\rho\{h - \sqrt{[(h+z)^2 + a^2]} + \sqrt{(z^2 + a^2)}\} \quad \ldots\ldots\ldots\ldots(1),$$

where z is the distance of the point from the centre of the nearer end.

Taking the centre of the plane face as origin and the axis of x through the particle and the mass of the particle as unity, its equation of motion is

$$\ddot{x} = \frac{\partial V}{\partial x} \quad\dots\dots\dots\dots\dots\dots\dots(2),$$

where $\partial V/\partial x$ is the force of attraction at $(x, 0, 0)$. By MacLaurin's theorem

$$\frac{\partial V}{\partial x} = \left(\frac{\partial V}{\partial x}\right)_0 + x\left(\frac{\partial^2 V}{\partial x^2}\right)_0 + \text{higher powers}\dots\dots.$$

And at the origin $\left(\dfrac{\partial V}{\partial x}\right)_0 = 0$, by symmetry, and $\left(\dfrac{\partial^2 V}{\partial x^2}\right)_0 = \left(\dfrac{\partial^2 V}{\partial y^2}\right)_0$. But

$$\left(\frac{\partial^2 V}{\partial x^2}\right)_0 + \left(\frac{\partial^2 V}{\partial y^2}\right)_0 + \left(\frac{\partial^2 V}{\partial z^2}\right)_0 = 0;$$

therefore $\qquad \dfrac{\partial V}{\partial x} = x\left(\dfrac{\partial^2 V}{\partial x^2}\right)_0 + \dots = -\tfrac{1}{2}x\left(\dfrac{\partial^2 V}{\partial z^2}\right)_0 + \dots \quad\dots\dots\dots(3),$

where the expression (1) represents $-\partial V/\partial z$ at $(0, 0, z)$. Hence, on the axis

$$\frac{\partial^2 V}{\partial z^2} = 2\pi\mu\rho\left\{\frac{h+z}{\sqrt{[(h+z)^2+a^2]}} - \frac{z}{\sqrt{(z^2+a^2)}}\right\},$$

and at the origin $\qquad \left(\dfrac{\partial^2 V}{\partial z^2}\right)_0 = \dfrac{2\pi\mu\rho h}{\sqrt{(h^2+a^2)}} \quad\dots\dots\dots\dots\dots(4).$

Hence, from (2), (3) and (4), the equation of motion is

$$\ddot{x} + \frac{\pi\mu\rho h}{\sqrt{(h^2+a^2)}}\, x = 0,$$

giving oscillations of the period stated.

(ii) *A solid bounded by the plane xy and the half of the ellipsoid* $x^2/a^2 + y^2/b^2 + z^2/c^2 = 1$ *on the positive side of the plane has a volume density* $\mu z(1/a^2 + 1/b^2 + 3/c^2)$. *Prove that it produces the same field of attraction outside itself as would a layer of surface density*

$$\tfrac{1}{2}\mu\left(1 - x^2/a^2 - y^2/b^2\right)$$

over the plane boundary and $\mu z/p$ *over the curved surface, when p is the central perpendicular on the tangent plane.* [C. 1928]

To prove the equivalence of the two fields of attraction in external space, it is only necessary to shew that if we reverse the sign of one distribution then the two distributions together produce a zero field at external points.

The potential in this combined field may therefore be assumed to be zero outside the solid and some function inside the solid which vanishes over the surface. We therefore write

$$V = Az\left(1 - x^2/a^2 - y^2/b^2 - z^2/c^2\right), \quad \text{inside,}$$

and $\qquad\qquad V' = 0, \quad \text{outside.}$

Hence we get

$$\frac{\partial V}{\partial x} = -\frac{2Azx}{a^2}, \quad \frac{\partial V}{\partial y} = -\frac{2Azy}{b^2}, \quad \frac{\partial V}{\partial z} = A\left(1 - \frac{x^2}{a^2} - \frac{y^2}{b^2} - \frac{3z^2}{c^2}\right)$$

and

$$\frac{\partial^2 V}{\partial x^2} = -\frac{2Az}{a^2}, \quad \frac{\partial^2 V}{\partial y^2} = -\frac{2Az}{b^2}, \quad \frac{\partial^2 V}{\partial z^2} = -\frac{6Az}{c^2}.$$

If ρ is the volume density, we have

$$-4\pi\rho = \nabla^2 V = -2Az\left(\frac{1}{a^2} + \frac{1}{b^2} + \frac{3}{c^2}\right)$$

or, putting $A = 2\mu\pi$, $\rho = \mu z\left(\dfrac{1}{a^2} + \dfrac{1}{b^2} + \dfrac{3}{c^2}\right)$ inside,

and $\rho = -\dfrac{1}{4\pi}\nabla^2 V' = 0$ outside $\Bigg\}$ ....(1).

Again differentiating along the outward normal, on the plane boundary we have

$$4\pi\sigma = \left(-\frac{\partial V}{\partial z} + \frac{\partial V'}{\partial z}\right)_{z=0}$$

$$= -A\left(1 - \frac{x^2}{a^2} - \frac{y^2}{b^2}\right)$$

and, putting $A = 2\mu\pi$,

$$\sigma = -\tfrac{1}{2}\mu\left(1 - \frac{x^2}{a^2} - \frac{y^2}{b^2}\right) \quad \text{......................(2).}$$

Also, on the curved boundary

$$4\pi\sigma = \frac{\partial V}{\partial n} - \frac{\partial V'}{\partial n}$$

$$= p\left(\frac{x}{a^2}\frac{\partial V}{\partial x} + \frac{y}{b^2}\frac{\partial V}{\partial y} + \frac{z}{c^2}\frac{\partial V}{\partial z}\right),$$

which reduces to $-2Az/p$, so that, putting $A = 2\mu\pi$, we have

$$\sigma = -\mu z/p \quad \text{..............................(3).}$$

Since the distributions (1), (2), (3) give a zero field outside the body, it follows that when the sign of the surface distribution is changed it produces the same external field as the volume distribution.

EXAMPLES

1. If the potential in the interior of an attracting mass bounded by two concentric spherical surfaces be proportional to $1/\sqrt{r}$, where r is the distance from the centre, find the law of density, and the potential exterior to the attracting matter.

Shew that at any point of space where the density is ρ and potential V, the magnitude of the force is given by the expression

$$\frac{1}{\sqrt{2}}\{8\pi\rho V + \nabla^2 V^2\}^{\frac{1}{4}}. \qquad \text{[C. 1892]}$$

2. Find the distribution of matter which produces a potential $\mu(x^2+y^2+z^2-2ax)$ at all points inside the sphere $x^2+y^2+z^2-2ax=0$ and zero potential at all points outside it. [C. 1908]

3. Find the distribution of matter, none of which is external to the sphere $r=a$, which produces a potential $-2\pi\rho x^2$ at points within or on the sphere, the origin being the centre, x the abscissa, and ρ a constant. [C. 1914]

4. What distribution of matter will produce potential

$$\frac{M}{a}\left(1-\frac{x}{3a}\right) \quad \text{or} \quad \frac{M}{r}\left(1-\frac{ax}{3r^2}\right),$$

according as r is less or greater than a? [C. 1902]

5. The potential outside an infinite elliptic cylinder

$$x^2/a^2+y^2/b^2-1=0$$

is zero; inside it is $V=x^2/a^2+y^2/b^2-1$. Find the distribution of matter. [C. 1915]

6. If the potential within an attracting medium having spherical symmetry is given by $V=(c^2+r^2)^{-\frac{1}{2}}$, where r is the distance from the centre and c a constant, prove that the density is given by

$$\rho=3c^2/4\pi(c^2+r^2)^{\frac{5}{2}},$$

and that the potential energy of the whole medium is $-3\pi/32c$. [M. T. 1915]

7. Find the distribution of matter that gives rise to potentials

$$V_1=\tfrac{4}{3}\mu\pi a^3+\tfrac{4}{15}\mu\pi a(2x^2-y^2-z^2), \quad r<a$$

and

$$V_2=\tfrac{4}{3}\mu\pi\frac{a^4}{r}+\tfrac{4}{15}\mu\pi\frac{a^6}{r^5}(2x^2-y^2-z^2), \quad r>a.$$

8. Prove (by taking $x^4+y^4+z^4-1$ as a potential function) that, if the space enclosed by the surface $x^4+y^4+z^4=1$ is filled with attracting matter of density proportional to the square of the distance from the origin, its attraction at all outside points will be the same as if its mass were distributed over the boundary surface with surface density inversely proportional to the distance of the tangent plane from the origin. [M. T. 1890]

9. Shew that the family of surfaces defined by

$$x^2+y^2=\text{const.}$$

can be a family of equipotential surfaces in free space, and find the law of potential. [C. 1906]

10. Shew that the system of co-axial cylinders

$$x^2+y^2+2\lambda x+c^2=0$$

can form a system of equipotential surfaces. [London Univ. 1934]

11. Shew that the cylinders $x^2 + y^2 = 2\lambda x$ are a possible set of equipotential surfaces in empty space, but that the spheres $x^2 + y^2 + z^2 = 2\lambda x$ are not.

In the former case find the potential as a function of λ.

[London Univ.]

12. Two equal circular cylinders, with their axes in the same straight line and their centres of gravity at a great distance r apart, attract one another according to the law of the inverse square of the distance. Shew that the resultant attraction is approximately

$$M^2 r^{-2} + 6M (A - C) r^{-4},$$

where M is the mass of one of the cylinders, C is its moment of inertia about its axis and A its moment of inertia about a perpendicular line through the centre of gravity.

13. If two distributions of matter, A and B, have throughout a region T, to which both are external, the same equipotential surfaces, the attractions due to A and B throughout the region T are in the same direction and in a constant ratio.

14. If two distributions of matter, A and B, have the same closed surface S, enclosing both, as an equipotential surface, every surface outside S which is an equipotential surface for A is also an equipotential surface for B.

---

15. A finite solid of revolution is cut at right angles by a plane, and a particle is placed on the plane, a smooth surface, near to its centre O. Shew that the time of a small oscillation is

$$2\pi \Big/ \left(\frac{1}{2}\frac{\partial^2 V}{\partial z_0{}^2}\right)^{\frac{1}{2}},$$

where V is the potential of the solid, the axis of the solid is the axis of z and $\partial^2 V/\partial z_0{}^2$ denotes the value at O.

16. A gravitating solid of revolution is cut by a plane perpendicular to the axis. A particle is fastened by a fine string of length l to a point in the prolongation of the axis, so that when the string is vertical the particle just does not touch the plane face at its centre O. Assuming the conditions such that when the particle is slightly disturbed the motion is that of a simple pendulum, shew that the time of a small oscillation is

$$2\pi \left(\frac{l}{R + \frac{1}{2}lR'}\right)^{\frac{1}{2}},$$

where R is the force exerted by the solid on unit mass at O and R' is the derivative of that force at O, taken outside the solid, along the axis. [C. 1892]

17. A uniform circular disc is of mass M and radius a and has its centre of gravity at O. A particle of mass m is situated at P, at a great distance r from O, and OP makes an angle θ with the axis of the disc. If the attractive force of the particle on the disc is reduced to a force through O and a couple, shew that the moment of the couple is $(\frac{3}{8}\gamma mMa^2 \sin 2\theta)/r^3$, and find the magnitude of the force.

[London Univ. 1938]

18. Prove that the x-component of attraction of a gravitating medium on a small sphere of itself with its centre at (x, y, z) is approximately

$$-\frac{1}{3\gamma} a^3 \frac{\partial V}{\partial x} \nabla^2 V - \frac{1}{30\gamma} a^5 \nabla^2 \left(\frac{\partial V}{\partial x} \nabla^2 V \right),$$

where a is the radius of the sphere and V the potential at (x, y, z).

[C. 1932]

19. Shew that, if the curves $f(x, y, \lambda) = 0$ form a system of equipotential lines in free space for a two-dimensional system, the surfaces formed by their revolution round the axis of x cannot be a system of equipotential surfaces in free space unless

$$\frac{1}{y} \frac{\partial \lambda}{\partial y} \Big/ \left\{ \left(\frac{\partial \lambda}{\partial x} \right)^2 + \left(\frac{\partial \lambda}{\partial y} \right)^2 \right\}$$

is a constant or a function of λ.

[C. 1905]

20. Shew that the family of cylinders

$$(x^2 + y^2)^3 - 2a^3 (x^3 - 3xy^2) + a^6 = \text{const.}$$

is a possible form of equipotential surfaces, and find the corresponding potential.

[London Univ. 1938]

21. Shew that $(x - c)^2 + y^2 = \lambda [(x + c)^2 + y^2]$ represents a family of equipotential surfaces, and that

$$V = A \log \lambda + B,$$

where V is the potential function and A and B are arbitrary constants.

[London Univ. 1939]

22. Determine a distribution of matter whose potential shall be constant over confocal elliptic cylinders. [C. 1915]

23. Shew that the family of ellipsoids

$$\frac{x^2}{a + \lambda} + \frac{y^2}{b + \lambda} + \frac{z^2}{c + \lambda} = 1,$$

where λ is a variable parameter and a, b, c are constants, is a possible form of equipotential surfaces, and express the potential in terms of a, b, c and λ.

24. The potential functions in the three regions bounded by the spheres $r = b$ and $r = a\,(>b)$ are

$$V_1 = \tfrac{4}{3}\pi\gamma\rho\left[\frac{2a^3 - b^3}{r} + \frac{a^5}{5r^5}(2z^2 - x^2 - y^2)\right], \quad r > a;$$

$$V_2 = \tfrac{4}{3}\pi\gamma\rho\left[\tfrac{5}{2}a^2 - \tfrac{1}{2}r^2 - \frac{b^3}{r} + \tfrac{1}{5}(2z^2 - x^2 - y^2)\right], \quad a > r > b;$$

and $\quad V_3 = \tfrac{4}{3}\pi\gamma\rho\,[\tfrac{5}{2}a^2 - \tfrac{3}{2}b^2 + \tfrac{1}{5}(2z^2 - x^2 - y^2)], \quad b > r > 0.$

Find the volume and surface distributions of matter and the total mass of the distribution. [London Univ. 1939]

25. In the interior of a spherical distribution of matter of radius a the gravitational potential at a distance r from the centre is

$$\frac{A\sin(\pi r/a)}{\pi r/a} + \text{const.}$$

Prove that the density at any point is $\dfrac{\pi A}{4Ga^2}\dfrac{\sin(\pi r/a)}{\pi r/a}$, and that the mass of the whole distribution is Aa/G, G being the constant of gravitation.

Prove also that the negative potential energy (compared with a state of infinite diffusion) is $\tfrac{3}{4}A^2a/G$. [M. T. 1925]

26. Find the attracting systems whose potential is

$$\tfrac{2}{63}\pi xyz\,(9a^2 - 7r^2) + 4\pi rx/(x^2 + y^2) \quad \text{for} \quad r < a,$$

and $\quad\quad \tfrac{4}{63}\pi xyz\dfrac{a^9}{r^7} + 4\pi ax/(x^2 + y^2) \quad \text{for} \quad r > a,$

where $r^2 = x^2 + y^2 + z^2$. [C. 1927]

27. The axes being rectangular, determine what volume and surface distributions of matter will give rise to potential $Kxyz\,(a - x - y - z)$, where K and a are constants, at all points within the tetrahedron bounded by the co-ordinate planes and the plane $x + y + z = a$, and potential zero at all points outside the tetrahedron.

Shew that the total mass of the matter contained within the tetrahedron is $Ka^5/80\pi\gamma$. [London Univ. 1938]

28. An infinite uniform straight line occupying the positive part of the axis of z attracts according to the law of nature. Shew from elementary considerations that its potential is a function of $r - z$, where r is distance from the origin, and hence prove that its potential is

$$c\log(r - z) + c'.$$

29. Prove that, if the matter be distributed symmetrically about the axis Oz, $\partial^2 V/\partial z\,\partial r = 0$, at every point of this axis, where r is the distance from this axis. Prove also that if the origin be a point of equilibrium, the lines of force near it are given by the equation

$$zr^2 = \text{const.} \tag{C. 1901}$$

30. Compare roughly the couples exerted on the earth by the sun and moon respectively, supposing them to be at the same angular distance from the equator and the orbits of the moon and earth to be circles. Take the ratio of the mass of the moon to that of the earth to be 1/80.
[M. T. 1922]

31. Obtain the mutual potential energy of two gravitating bodies placed at a considerable distance from one another, neglecting powers of the inverse distance higher than the third, and obtain the specification of the moment, about the centre of mass of either body, of the attraction exercised upon that body by the other.

Shew that the couples exercised by the sun and the moon on the earth neutralize one another at an instant when they are on the same meridian if

$$\frac{\sin 2\omega'}{\sin 2\omega} = -\frac{S(E+M)}{M(E+S)}\left(\frac{T'}{T}\right)^2,$$

where ω' and ω are the angular distances of moon and sun from the equator, E, M and S are the masses of the earth, moon and sun, T' and T being the lengths of the sidereal month and year respectively. The orbits of the moon and earth may be considered circular.
[M. T. 1937]

32. O is the centre of a uniform cube of mass M and P is an external point. The distance OP is large compared with the edge a of the cube, and the direction cosines of OP referred to three concurrent edges of the cube are l, m, n. Shew how to express the gravitational potential at P in a power series in OP^{-1}, and evaluate the potential in terms of OP, M, l, m, n and a, rejecting all terms in OP^{-1} of higher order than the fifth.
[P. 1936]

33. Shew that the equipotential surfaces of a uniform equilateral triangular lamina (of side $2a$), which are at a great distance from the lamina, approximate to the surfaces of revolution obtained by rotating the curves

$$\frac{1}{r} + \frac{a^2(1 - 3\cos^2\theta)}{12r^3} = \text{const.}$$

about the initial line.
[London Univ. 1938]

34. Prove that, in the case of two equal cubes at a great distance apart, the third order term in the expression for their mutual gravitational potential energy disappears and obtain the next non-vanishing term.
[M. T. 1929]

35. Prove that for two gravitating solids, if one is a fixed sphere and the other has unequal principal moments of inertia and is free to turn about its centre of gravity, the only positions of equilibrium are those in which a principal axis at the centre of gravity of the second body is directed towards the centre of the sphere.
[P. 1933]

36. Prove that a distribution of matter of surface density $mp/4\pi r^3$ over an ellipsoid will produce the same potential at every external point as a homogeneous concentric sphere of mass m within the ellipsoid, r and p being the central radius and the perpendicular from centre on the tangent plane. What is the potential of this distribution inside the ellipsoid? [C. 1892]

37. In a region of space which does not intersect the plane $y = 0$ a function ϕ satisfies $\nabla^2\phi = 0$, and the surfaces with the equations $\phi = $ constant are portions of the elliptical cylinders

$$\frac{x^2}{a^2+\lambda}+\frac{y^2}{\lambda}=1,$$

where λ is a variable parameter. By transforming to co-ordinates ξ, η, ζ defined by

$$x = a\cosh\xi\cos\eta, \quad y = a\sinh\xi\sin\eta, \quad z = \zeta,$$

$$\xi \geqslant 0, \quad 2\pi > \eta \geqslant 0,$$

prove that ϕ is of the form $A\xi + B$, where A and B are constants.
[P. 1937]

Chapter V

GREEN'S THEOREM

5·1. In **1·3** and **1·31** the following theorem was established:
If V, V' are functions which with their first derivatives are finite, continuous, and single-valued through a singly-connected region* bounded by a surface S,

$$\int\left(\frac{\partial V}{\partial x}\frac{\partial V'}{\partial x}+\frac{\partial V}{\partial y}\frac{\partial V'}{\partial y}+\frac{\partial V}{\partial z}\frac{\partial V'}{\partial z}\right)dv$$

$$=\int V\frac{\partial V'}{\partial n}dS-\int V\nabla^2 V'\,dv$$

$$=\int V'\frac{\partial V}{\partial n}dS-\int V'\nabla^2 V\,dv \quad \ldots\ldots(1),$$

where the surface integrals are over the boundary S and the volume integrals throughout the bounded region and $\partial/\partial n$ denotes differentiation in the direction of the outward normal.

The proof in **1·3** was for a region bounded by a single surface, but it requires little consideration to see that it is also true for a region with several boundaries, such as that bounded externally by S and internally by S_1 and S_2 as in the figure. But in this case the surface integrals must extend to the three surfaces S, S_1 and S_2, and care must be taken that the normal differentiation is outwards from the region of volume integration over each surface, as indicated by the arrows. The theorem needs

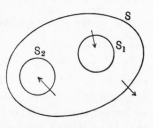

---

* A singly-connected region is one in which every closed curve can be contracted to a point without passing out of the region. The space between two concentric spheres is singly-connected, but the space inside an anchor ring is doubly-connected.

modification for multiple-valued functions and multiply-connected space,* but we are not concerned here with such modifications.

In the applications of this theorem which we shall consider, V and V' will be potential functions of attracting masses. We know that the potential function V of a system of attracting particles becomes infinite at each particle, if we regard the particle as a concentration of matter at a point; also, that there is a discontinuity in the normal derivative of V across any surface on which there is a surface density of matter (3·7), consequently such points and surfaces of discontinuity must be excluded from the region of integration in applications of the theorem. But with this proviso the region might be regarded as bounded externally by an infinite sphere and internally by one or more surfaces adapted to the particular problem under consideration. In such a case integration over the infinite sphere must be included among the surface integrals in (1). But if V, V' are potentials of finite masses within a finite region, and R denotes the radius of a large sphere, then on this sphere V, V' are each of order $1/R$; $\partial V/\partial n$, $\partial V'/\partial n$ are of order $1/R^2$ and dS is of order $R^2 d\omega$, where $d\omega$ is a solid angle. Hence $\int V \dfrac{\partial V'}{\partial n} dS$ is of order $1/R$ and may be neglected when R tends to infinity. Similarly for $\int V' \dfrac{\partial V}{\partial n} dS$, so that when V, V' are potentials of finite masses the surface integrals over the infinite sphere vanish.

It is evident that very varied applications of Green's theorem can be made, but in every case the first essential is to specify to what region of space the theorem is applied and to ensure that V, V' and their derivatives are continuous in the way required within that region.

5·2. Applications. Gauss's Theorem. Put $V' = 1$, and let the region be bounded externally by a single surface S, and suppose it to enclose matter of finite volume density ρ, where

---

* See the author's *Hydrodynamics*, 4·81.

ρ may be zero through any part or parts of the region. The theorem gives

$$\int \frac{\partial V}{\partial n} dS = \int \nabla^2 V dv \quad \dots\dots\dots\dots(1).$$

But, by Poisson's equation, $\nabla^2 V = -4\pi\rho$, wherever the matter exists; therefore

$$\int \frac{\partial V}{\partial n} dS = -4\pi \int \rho \, dv,$$

or $\qquad \int R_n dS = -4\pi \quad$ (mass enclosed by S) $\quad \dots\dots(2);$

which is Gauss's theorem.

It is instructive to see how the application of Green's theorem needs modification when the surface S encloses not only matter of finite volume density ρ, but also material particles, implying points at which V is infinite, and surfaces on which there is a surface density, implying discontinuity in $\partial V/\partial n$.

It will be sufficient for our purpose to suppose that the surface S encloses a body of mass M whose density at any point is ρ, a particle of mass m at a point P, and a surface S' on which there is matter of surface density σ. In order to apply Green's theorem to the region enclosed by S, we must exclude the point P and the surface S' from the range of integration. This is done by drawing a sphere Σ of centre P and small radius ϵ, and by drawing surfaces parallel to S' and close together on opposite

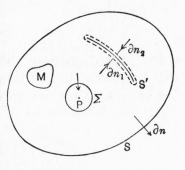

sides of S' as shewn in the figure. If these surfaces are sufficiently close together, end effects may be ignored. We now apply Green's theorem in the form (1) above to the region bounded externally by S and internally by Σ and the surface which surrounds S'. We have

$$\int_S \frac{\partial V}{\partial n} dS + \int_\Sigma \frac{\partial V}{\partial n} d\Sigma + \int_{S'} \left(\frac{\partial V}{\partial n_1} + \frac{\partial V}{\partial n_2} \right) dS' = \int \nabla^2 V dv \dots(3),$$

where in the second integral ∂n is directed out of 'the region', i.e. into the small sphere, and in the third integral ∂n_1 and ∂n_2 are both directed towards S'.

In the immediate neighbourhood of P the part of V which depends on the particle m at P is the only part which will contribute to the surface integral, if ϵ is taken small enough; and the second integral is therefore

$$\int \frac{m}{\epsilon^2}\,\epsilon^2 d\omega = 4\pi m.$$

Then in the third integral $\partial V/\partial n_1$ and $\partial V/\partial n_2$ are the attraction components normal to S', and by comparing with **3·7** (1) and paying due regard to sign we see that $\dfrac{\partial V}{\partial n_1} + \dfrac{\partial V}{\partial n_2} = 4\pi\sigma$, so that the third integral is $4\pi\int\sigma dS'$. Hence (3) is equivalent to

$$\int_S \frac{\partial V}{\partial n}\,dS + 4\pi m + 4\pi\int\sigma dS' = -4\pi\int\rho\,dv,$$

or
$$\int_S R_n\,dS = -4\pi M - 4\pi m - 4\pi\int\sigma dS'$$

$$= -4\pi \quad \text{(mass enclosed by } S\text{)}.$$

5·21. Green's Reciprocal Theorem. Let V, V' be the potentials of two different distributions of matter of finite volume densities ρ, ρ' situated anywhere in the finite region of space. Apply Green's theorem **5·1** (1) in the form

$$\int\left(V\frac{\partial V'}{\partial n} - V'\frac{\partial V}{\partial n}\right)dS = \int(V\nabla^2 V' - V'\nabla^2 V)\,dv \quad \text{...(1)}$$

to the region bounded by the infinite sphere. As explained in **5·1** the surface integral over the infinite sphere vanishes, and since by Poisson's equation $\nabla^2 V = -4\pi\rho$ and $\nabla^2 V' = -4\pi\rho'$, therefore (1) reduces to

$$\int V\rho'\,dv = \int V'\rho\,dv \dots\dots\dots\dots\dots\dots(2).$$

This is the reciprocal relation obtained in **3·8**, either side of the equality representing the mutual work of the two distributions of matter.

We might infer that if the distributions of matter include separate attracting particles and surfaces on which matter is condensed, then the complete expression of the theorem will be a relation

$$\int V\rho'\,dv + \Sigma\,(Vm') + \Sigma\int V\sigma'\,dS'$$

$$= \int V'\rho\,dv + \Sigma\,(V'm) + \Sigma\int V'\sigma\,dS \quad \ldots\ldots(3),$$

where m, m' denote the masses of typical particles of the two distributions and $\sigma\,dS$, $\sigma'\,dS'$ elements of the surface distributions and the Σ's imply summation for all such particles and surfaces. But though this is a true result, the foregoing argument does not constitute a proof of it, because the region of integration in (1) (all space) now includes the particles, i.e. points at which V or V' becomes infinite, and the material surfaces across which $\partial V/\partial n$ or $\partial V'/\partial n$ is discontinuous, and we may not apply (1) without excluding such points and surfaces from the region of integration, so that the result (2) establishes nothing concerning the case of particles or material surfaces.

Let m_1, m_1' be the masses of typical particles of the two systems at A, A'. Surround them by spheres Σ_1, Σ_1' of small radii ϵ_1, ϵ_1' and centres A, A'. Let S, S' be typical surfaces with surface densities σ, σ' in the two systems and let them each be enclosed by parallel surfaces a small distance apart, as in 5·2. We can now apply

the theorem (1) to the region bounded externally by the infinite sphere and internally by the small spheres and other surfaces so as to exclude the particles and material surfaces from the region. The surface integral on the left in (1) must now be understood to refer to the infinite sphere and to all the inner boundaries, and, as before, over the infinite sphere it vanishes. Consider

$$\int_{\Sigma_1} \left(V \frac{\partial V'}{\partial n} - V' \frac{\partial V}{\partial n} \right) d\Sigma_1$$

over the small sphere Σ_1.

Since ϵ_1 is small, on Σ_1 the principal part of V is m_1/ϵ_1, so that we may put $\dfrac{m_1}{\epsilon_1} + \eta$ for V, where η is finite; and, taking account of the sign of ∂n, we may put $\dfrac{m_1}{\epsilon_1^{\,2}} + \eta_1$ for $\dfrac{\partial V}{\partial n}$, where η_1 is finite; and $d\Sigma_1 = \epsilon_1^{\,2} d\omega$, so that the integral is

$$\int \left\{ \left(\frac{m_1}{\epsilon_1} + \eta \right) \frac{\partial V'}{\partial n} - V' \left(\frac{m_1}{\epsilon_1^{\,2}} + \eta_1 \right) \right\} \epsilon_1^{\,2} d\omega,$$

and if we make $\epsilon_1 \to 0$, this reduces to $-4\pi V_A' m_1$, where V_A' is the value of V' at A. In like manner the same integral over the small sphere Σ_2 reduces to $4\pi V_{A'} m_1'$.

Taking the same integral over the surfaces which surround S (neglecting edge effects), we have

$$\int_S \left\{ V \left(\frac{\partial V'}{\partial n_1} + \frac{\partial V'}{\partial n_2} \right) - V' \left(\frac{\partial V}{\partial n_1} + \frac{\partial V}{\partial n_2} \right) \right\} dS,$$

and taking account of the signs of ∂n_1, ∂n_2, we see that by 3·7 $\dfrac{\partial V}{\partial n_1} + \dfrac{\partial V}{\partial n_2} = 4\pi\sigma$; and $\dfrac{\partial V'}{\partial n_1} + \dfrac{\partial V'}{\partial n_2} = 0$ since by hypothesis none of the second distribution of matter resides on S and V' is continuous across S. The integral therefore reduces to $-4\pi \int V'\sigma dS$. In like manner the similar integral over the surfaces surrounding S' reduces to $4\pi \int V\sigma' dS'$.

Then substituting from Poisson's equation on the right-hand side of (1), dropping the factor 4π, collecting like terms, and summing for all particles and material surfaces, we obtain the result (3).

We conclude that whenever we use Green's theorem to prove something about attracting particles or surfaces it is not through the volume integrals but through the surface integrals that we must expect their masses to enter into the result.

5·3. Potential energy.

Let V be the potential due to a system of particles, surfaces on which matter is condensed and finite bodies. It will suffice for our purpose to consider a single particle of mass m at a point P, a single surface S' on which there is matter of surface density σ, and let the finite bodies be represented by a distribution of volume density ρ. Take the infinite sphere as the external boundary of the region and the same internal boundaries as in 5·2, excluding from the region the point P and the surface S'.

Apply Green's theorem in the form 5·1 (1), putting $V' = V$. For the reason given in 5·1 the integral over the infinite sphere vanishes, so that we have in this case

$$\int \left\{ \left(\frac{\partial V}{\partial x}\right)^2 + \left(\frac{\partial V}{\partial y}\right)^2 + \left(\frac{\partial V}{\partial z}\right)^2 \right\} dv$$

$$= \int_\Sigma V \frac{\partial V}{\partial n} d\Sigma + \int_{S'} V \left(\frac{\partial V}{\partial n_1} + \frac{\partial V}{\partial n_2}\right) dS' - \int V \nabla^2 V dv \dots (1),$$

where the volume integrals are through all space, save the vanishing regions which we have excluded, and the surface integrals are over the small sphere round P and over the surface S' as in 5·2.

The integral on the left is $\int R^2 dv$, where R is the resultant force at a point of the element dv.

The first integral on the right, as in 5·2, is $\int V \frac{m}{\epsilon^2} \epsilon^2 d\omega$, where if ϵ is small enough we may take for V its value at P, say V_P, so that the integral is $4\pi m V_P$.

The second integral on the right, as in 5·2, is $4\pi \int_{S'} V\sigma dS'$, and the third integral on the right is $4\pi \int V\rho dv$, integrated wherever there is a volume density of matter.

Therefore

$$\int R^2 dv = 4\pi m V_P + 4\pi \int_{S'} V \sigma dS' + 4\pi \int V \rho dv \ldots \ldots (2).$$

It is clear that the presence of other attracting particles and surfaces would contribute other terms like the first and second on the right.

But from **3·81** the potential energy of this attracting system, there denoted by $-\frac{1}{2}\int V dm$, is in this case

$$-\frac{1}{2}\left(m V_P + \int V \sigma dS' + \int V \rho dv \right),$$

and, from (2), this is equal to $-\dfrac{1}{8\pi}\int R^2 dv$ integrated through all space. It follows that the potential energy of any attracting masses may be represented by an energy density of amount $- R^2/8\pi$ distributed through all space.

5·31. *If V is a solution of Laplace's equation which with its first and second derivatives is single valued and finite within a region bounded by a closed surface S, and $\partial V/\partial n = 0$ over the surface, then V is constant throughout the region.*

In **5·1** (1) put $V' = V$, so that

$$\int \left\{ \left(\frac{\partial V}{\partial x}\right)^2 + \left(\frac{\partial V}{\partial y}\right)^2 + \left(\frac{\partial V}{\partial z}\right)^2 \right\} dv = \int V \frac{\partial V}{\partial n} dS - \int V \nabla^2 V dv \ \ldots(1).$$

Since $\partial V/\partial n = 0$ over S, and $\nabla^2 V = 0$ inside S, therefore the integral on the left of (1) is equal to zero. This requires that, at every point of the region, $\dfrac{\partial V}{\partial x} = \dfrac{\partial V}{\partial y} = \dfrac{\partial V}{\partial z} = 0$, so that V must be constant throughout the region.

It follows that if V_1, V_2 are two solutions of Laplace's equation satisfying the above conditions within the bounded region and making $\partial V_1/\partial n = \partial V_2/\partial n$ over S, then $V_1 - V_2$ also satisfies the same conditions and makes $\partial (V_1 - V_2)/\partial n = 0$ over S, so that $V_1 - V_2$ must be constant throughout the region.

5·4. Green's equivalent layer. Let S be a closed equi-potential surface of a given distribution of attracting matter of which some or all lies inside S. For the sake of brevity we assume that the matter inside S can be represented by a volume density ρ. Let r denote distance from a point Q *outside* S. Apply Green's theorem to the region inside S in the form

$$\int V \frac{\partial V'}{\partial n} dS - \int V \nabla^2 V' dv = \int V' \frac{\partial V}{\partial n} dS - \int V' \nabla^2 V dv \quad ...(1),$$

where V denotes the potential of the whole distribution of matter and $V' = 1/r$.

This gives

$$\int V \frac{\partial (1/r)}{\partial n} dS - \int V \nabla^2 \frac{1}{r} dv = \int \frac{1}{r} \frac{\partial V}{\partial n} dS - \int \frac{1}{r} \nabla^2 V dv \quad ...(2).$$

Let V_s be the value of V on the equipotential surface S, then the first integral in (2) is $V_s \int \frac{\partial (1/r)}{\partial n} dS$. But $1/r$ is the potential due to a particle of unit mass at Q, and Q is outside S, so that by Gauss's theorem $\int \frac{\partial (1/r)}{\partial n} dS = 0$; therefore the first integral vanishes. The second integral also vanishes since $\nabla^2 \frac{1}{r} = 0$ throughout the range of integration. Also $\nabla^2 V = -4\pi\rho$ throughout the same range, so that (2) reduces to

$$0 = \int \frac{1}{r} \frac{\partial V}{\partial n} dS + 4\pi \int \frac{\rho dv}{r}.$$

But $\int \frac{\rho dv}{r}$ is the potential at Q due to the matter *inside* S, so that denoting this by $(V)_Q$ we have

$$(V)_Q = -\frac{1}{4\pi} \int \frac{1}{r} \frac{\partial V}{\partial n} dS \quad ....(3).$$

Therefore the part of the potential at any point *outside* S which is due to the matter *inside* S is the same as would be produced by a layer of matter of surface density

$$\sigma = -\frac{1}{4\pi}\frac{\partial V}{\partial n} \quad \dots\dots\dots\dots\dots\dots(4)$$

spread over S, where V is the potential due to the *whole* of the attracting matter.

Further, the amount of matter in the layer is given by

$$\int \sigma \, dS = -\frac{1}{4\pi}\int \frac{\partial V}{\partial n} \, dS$$

$$= -\frac{1}{4\pi}\int \nabla^2 V dv, \quad \textbf{5·2 (1)},$$

$$= \int \rho \, dv$$

$$= \text{mass contained within } S.$$

We have therefore proved that the matter contained within any closed equipotential surface S in a given field can be spread over the surface S, with a surface density $-\frac{1}{4\pi}(\partial V/\partial n)$ at any point, without altering the potential at any point of the field outside S. This distribution is known as **Green's equivalent layer.**

If we suppose that this distribution over the surface S of the matter inside S has been effected, S is now an equipotential surface containing no matter, so that the potential at any point inside S is now V_s the value on the surface [**4·3** (ii)].

This can also be established analytically thus: let r' be the distance from a point Q' *inside* S and let ρ' denote the density of the matter *outside* S. We apply Green's theorem in the form (1) to the region between S and the infinite sphere, taking V as before to be the potential of the whole distribution of matter and $V' = 1/r'$. The normal derivatives in (1) are now directed along inward normals to dS, so if ∂n is to retain the same meaning as in (3) and (4) we must change the signs of the surface integrals and write

$$-\int V\frac{\partial(1/r')}{\partial n}\,dS - \int V\nabla^2\frac{1}{r'}\,dv = -\int\frac{1}{r'}\frac{\partial V}{\partial n}\,dS - \int\frac{1}{r'}\nabla^2 V dv \dots(5).$$

The first term is $-V_s\int\dfrac{\partial(1/r')}{\partial n}\,dS$, and since r' is measured from a point Q' inside S, $\int\dfrac{\partial(1/r')}{\partial n}\,dS = -4\pi$, (Gauss), and the first term in (5) is

$4\pi V_s$. The second term vanishes as before, and the fourth term is $4\pi \int \dfrac{\rho' dv}{r'}$; so that

$$4\pi V_s = -\int \frac{1}{r'}\frac{\partial V}{\partial n}\, dS + 4\pi \int \frac{\rho' dv}{r'}.$$

Or, if we put σ for the surface density of Green's layer, as in (4),

$$\int \frac{\sigma dS}{r'} + \int \frac{\rho' dv}{r'} = V_s \quad \dots\dots\dots\dots\dots\dots(6).$$

We thus verify that the potential inside S due to the joint effects of Green's layer and the original matter outside S is the constant potential of the equipotential surface.

5·5. Dirichlet's problem.* *There exists a solution of Laplace's equation which with its first and second derivatives is single-valued, finite and continuous throughout a region bounded by a surface S and takes arbitrarily assigned values over the boundary S.*

Dirichlet's method of investigation is as follows: let V be any function which satisfies all the conditions save that it is not necessarily a solution of Laplace's equation, then we can prove that, of all such functions, that one which gives a minimum value to the integral

$$W = \int \left\{ \left(\frac{\partial V}{\partial x}\right)^2 + \left(\frac{\partial V}{\partial y}\right)^2 + \left(\frac{\partial V}{\partial z}\right)^2 \right\} dv$$

through the bounded region is a solution of Laplace's equation.

Let V_1 be the function which gives a minimum value to W. If V be any function which satisfies the same conditions of continuity and takes the same values over S, then $V = V_1 + V_2$, where $V_2 = 0$ on S. Conversely, let V_2 satisfy the same conditions of continuity, and let $V_2 = 0$ on S. If ϵ is any constant, then $V_1 + \epsilon V_2$ takes the assigned values on S and satisfies the conditions of continuity. Thus the set of functions to be considered are comprised in the formula $V_1 + \epsilon V_2$, where $V_2 = 0$ on S. Then

$$W = \int \left\{ \left(\frac{\partial V_1}{\partial x} + \epsilon \frac{\partial V_2}{\partial x}\right)^2 + \dots + \dots \right\} dv$$

$$= W_1 + 2\epsilon \int \left(\frac{\partial V_1}{\partial x}\frac{\partial V_2}{\partial x} + \dots\right) dv + \epsilon^2 W_2 \quad \dots\dots\dots(1).$$

---

* Peter Gustav Lejeune Dirichlet (1805–1859). German mathematician.

Since W_1 is the minimum value of W, $W - W_1$ must be positive whatever be the sign of ϵ, so that the second term on the right of (1) must vanish. But by Green's theorem the coefficient of 2ϵ

$$= \int V_2 \frac{\partial V_1}{\partial n} dS - \int V_2 \nabla^2 V_1 dv;$$

and $V_2 = 0$ on S, so that the first integral vanishes and therefore

$$\int V_2 \nabla^2 V_1 dv = 0$$

for all functions V_2. We must therefore have $\nabla^2 V_1 = 0$ throughout the bounded region; for if $\nabla^2 V_1$ had a sign throughout any portion of the region we could take V_2 to have the same sign and the integral would not vanish.

The remainder of the argument is an assertion that among the different functions which satisfy the conditions of continuity and take the assigned values over S there must be one which gives W a smaller value than the rest, and, from what precedes, this is a solution of Laplace's equation. But the argument is invalid, because Weierstrass has shewn that we are not entitled to assume the existence of a *minimum*, but only the existence of a lower limit to the possible values of W and we cannot assert that the limit is attained.

Many mathematicians have produced solutions of Dirichlet's problem but they are too lengthy for reproduction here.*

The theorem is also applicable to the region between the surface S and the infinite sphere if we impose the condition that the required function shall vanish like $1/r$ at infinity.

We are entitled to conclude from the argument given above that if there is a solution it is the only one. For the above argument shews that, if $\nabla^2 V_1 = 0$ and $V_2 = 0$ on S, then the coefficient of ϵ in (1) is zero, so that $W = W_1 + \epsilon^2 W_2$. Similarly if $\nabla^2 V = 0$, then since $V_1 = V - \epsilon V_2$, we have, by the same argument, $W_1 = W + \epsilon^2 W_2$. Hence $\epsilon^2 W_2 = 0$, and so $\epsilon V_2 = 0$, i.e. $V = V_1$ and the solution is unique.

* See H. Poincaré, *Théorie du Potentiel Newtonien*; and, for a bibliography of the subject, P. Appell, *Traité de Mécanique Rationelle*, t. III, p. 98.

5·6. *To find the solution of* $\nabla^2 V = 0$ *inside a region bounded by a closed surface S when the values of V and* $\partial V/\partial n$ *over S are known.*

Take Green's theorem in the form

$$\int \left(V \frac{\partial V'}{\partial n} - V' \frac{\partial V}{\partial n} \right) dS = \int (V \nabla^2 V' - V' \nabla^2 V) \, dv \quad \text{...(1)}.$$

Put $V' = 1/r$, where r is the radius
vector drawn from a point P inside S.
Take a sphere Σ of small radius ϵ
with P as centre and apply the
theorem (1) to the region bounded by
S and Σ. Then, since $\nabla^2 (1/r) = 0$ and
$\nabla^2 V = 0$ throughout the region, the
right-hand side of (1) is zero, and
therefore

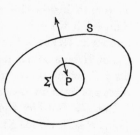

$$\int_S \left(V \frac{\partial (1/r)}{\partial n} - \frac{1}{r} \frac{\partial V}{\partial n} \right) dS + \int_\Sigma \left(V \frac{\partial (1/r)}{\partial n} - \frac{1}{r} \frac{\partial V}{\partial n} \right) d\Sigma = 0 \quad \text{...(2)}.$$

Taking account of the inward direction of ∂n over Σ we see
that $\partial (1/r)/\partial n = 1/\epsilon^2$, and since we may put $d\Sigma = \epsilon^2 d\omega$, we have
$\int_\Sigma V \frac{\partial (1/r)}{\partial n} d\Sigma = \int V d\omega$, and this being independent of ϵ we may
take ϵ as small as we please and write for the integral $4\pi V_P$,
where V_P denotes the value of V at P.

Again, $\int_\Sigma \frac{1}{r} \frac{\partial V}{\partial n} d\Sigma = \int_\Sigma \frac{1}{\epsilon} \frac{\partial V}{\partial n} \epsilon^2 d\omega \to 0$ as $\epsilon \to 0$; so that (2)
reduces to

$$V_P = \frac{1}{4\pi} \int_S \left(\frac{1}{r} \frac{\partial V}{\partial n} - V \frac{\partial (1/r)}{\partial n} \right) dS \quad \text{............(3)},$$

giving the value of V at any point inside S in terms of the
values of V and $\partial V/\partial n$ on the surface.

When the region is bounded by the surface S and the infinite
sphere, it is easy to shew that the same theorem holds good
provided we impose the condition that RV and $R^2 \partial V/\partial n$
remain finite as R tends to infinity, where R denotes distance
from an origin in the neighbourhood of S.

5·61. Green's function. We learn from Dirichlet's problem that there should be a unique solution V of Laplace's equation which with its first and second derivatives is finite through a bounded region and takes a prescribed form over the boundary, but the procedure of **5·6** seems to imply that we need to know not only the value of the function V but also the value of $\partial V/\partial n$ over the boundary in order to determine V at points inside the region. The problem can however be reduced if we can find a function of a more restricted form known as **Green's function** and defined as follows:

Let H be a solution of Laplace's equation inside the boundary S, which takes the value $-1/r$ on S, where r denotes the distance from an assigned point P inside S.

Let $G = H + 1/r$; then G is called *Green's function* for the point P and surface S; and by definition we see that G vanishes on S and satisfies Laplace's equation at every point inside S except at the point P where it becomes infinite.

In **5·6** (1) put $V' = H$, then since $\nabla^2 V = 0$ by hypothesis and $\nabla^2 H = 0$, we have

$$\int \left(V \frac{\partial H}{\partial n} - H \frac{\partial V}{\partial n} \right) dS = 0;$$

adding this, with a factor $-1/4\pi$, to the right-hand side of **5·6** (3), we get

$$V_P = \frac{1}{4\pi} \int \left\{ \left(H + \frac{1}{r} \right) \frac{\partial V}{\partial n} - V \frac{\partial}{\partial n} \left(H + \frac{1}{r} \right) \right\} dS$$

$$= \frac{1}{4\pi} \int \left(G \frac{\partial V}{\partial n} - V \frac{\partial G}{\partial n} \right) dS.$$

But G vanishes on S so that

$$V_P = -\frac{1}{4\pi} \int V \frac{\partial G}{\partial n} dS \quad \ldots\ldots\ldots\ldots\ldots(1).$$

By this means Dirichlet's problem is reduced to the determination of Green's function.

The case where the region is bounded by a closed surface S and the infinite sphere does not need special consideration if we impose like conditions as before to the way in which the functions vanish at infinity.

5·62. Green's function for a sphere. Let O be the centre of a sphere S of radius a and P an internal point at a distance f from O. Let P' be the inverse of P, so that $OP.OP' = a^2$. Let r, r' denote the distances of any point M from P and P'; and for a point M' on the sphere let $PM' = r_1$ and $P'M' = r_1'$.

Then by similar triangles $\dfrac{r_1'}{r_1} = \dfrac{OM'}{OP} = \dfrac{a}{f}$,

so that $\qquad\qquad\qquad\qquad \dfrac{a}{fr_1'} = \dfrac{1}{r_1}$ ....(1).

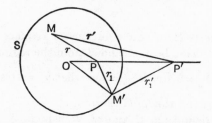

Now consider the function $H = -a/fr'$. It is a potential function due to a mass $(-a/f)$ placed at P', and therefore it satisfies Laplace's equation inside the sphere; and because of (1) it takes the form $-1/r$ on S.

Hence $\qquad\qquad\qquad\qquad G = \dfrac{1}{r} - \dfrac{a}{fr'}$ ....(2)

is the required Green's function for the sphere S and point P.

The same formula can be shewn to hold good when the point P lies outside the sphere at a distance f from the centre, P' then being inside and the symbols having the same meanings.

5·63. Reciprocal property of Green's function. *If P, Q are two points inside a region bounded by a surface S and $G(P, Q)$ denotes the value at Q of Green's function for the point P and surface S, then $G(P, Q) = G(Q, P)$.*

Apply Green's theorem in the form **5·6** (1) to the region bounded externally by S and internally by two spheres Σ_1, Σ_2 of small radii ϵ_1, ϵ_2 having their centres at P, Q.

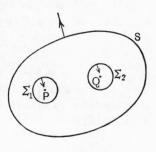

Put $V = G = \dfrac{1}{r} + H$, and $V' = G' = \dfrac{1}{r'} + H'$, where r, r' denote distances from P and Q and G, G' denote Green's functions for P and S and for Q and S respectively. Then we have

$$\int_S \left(G \frac{\partial G'}{\partial n} - G' \frac{\partial G}{\partial n} \right) dS + \int_{\Sigma_1} \left(G \frac{\partial G'}{\partial n} - G' \frac{\partial G}{\partial n} \right) d\Sigma_1$$

$$+ \int_{\Sigma_2} \left(G \frac{\partial G'}{\partial n} - G' \frac{\partial G}{\partial n} \right) d\Sigma_2 = \int (G \nabla^2 G' - G' \nabla^2 G) \, dv$$

$$\text{......(1)}.$$

By hypothesis H, H' are solutions of Laplace's equation, and since P, Q have been excluded from the region of volume integration $\nabla^2 \dfrac{1}{r} = 0$ and $\nabla^2 \dfrac{1}{r'} = 0$, so that $\nabla^2 G = 0$ and $\nabla^2 G' = 0$ and the right-hand side of (1) is zero.

Again, by definition, G and G' vanish on S, so that the first integral on the left is zero.

In the second integral on the left we put $d\Sigma_1 = \epsilon_1^2 \, d\omega$ so that

$$\int_{\Sigma_1} G \frac{\partial G'}{\partial n} \, d\Sigma_1 = \int \left(\frac{1}{\epsilon} + H \right) \frac{\partial G'}{\partial n} \, \epsilon^2 \, d\omega,$$

which clearly $\to 0$ as $\epsilon \to 0$, since H is assumed to be finite; and

$$-\int_{\Sigma_1} G' \frac{\partial G}{\partial n} \, d\Sigma_1 = -\int G_P' \left(\frac{1}{\epsilon^2} + \frac{\partial H}{\partial n} \right) \epsilon^2 \, d\omega,$$

and as $\epsilon \to 0$ this reduces to $-4\pi G_P'$, where G_P' denotes the value at P of Green's function for Q and S, i.e. $G(Q, P)$.

In the same way the third integral in (1) may be shewn to reduce to $4\pi G_Q$ or $4\pi G(P, Q)$; so that (1) reduces to

$$G(P, Q) = G(Q, P) \quad \text{.................(2)}.$$

The theorem is also true for a region bounded internally by a closed surface S, and externally by the infinite sphere, assuming that H and H' are at least of order R^{-1} at a great distance R.

EXAMPLES

1. Use Green's theorem to prove that, if V, V' are solutions of Laplace's equation such that $V = V'$ at all points of a closed surface S, then $V = V'$ throughout the interior of S.

2. Prove that, if V, V' are solutions of Laplace's equation such that $\partial V/\partial n = \partial V'/\partial n$ at all points of a closed surface S, then $V - V'$ is constant at all points inside S.

3. Shew that, if V_1, V_2 are the potentials of two equipotential surfaces which surround bodies of mass M which produce a gravitational field, the contribution of the space between these equipotential surfaces to the total energy of the field is $\frac{1}{2}M(V_1 - V_2)$.

4. Prove that there cannot be a function V which satisfies Laplace's equation inside a closed surface S and makes $\partial V/\partial n = -kV$ at all points of S, where ∂n denotes the normal drawn towards S from the inside and k is a positive number.

5. A number of bodies are formed of gravitating matter. Prove that the resultant force, resolved parallel to the axis of x, due to the attraction on one of the bodies, of uniform density ρ, is $\rho \int\int (x - \bar{x}) N \, dS$; where the integral is taken over the surface of the body, $\bar{x}$ is the abscissa of the centre of gravity of the body, and N is the normal attractive force at each point of the surface. [C. 1890]

6. Prove that, if dm, dm' denote elements of two different distributions of matter which have the same external level surfaces, and ϕ any function which satisfies Laplace's equation, then

$$\int \phi \, dm : \int \phi \, dm' = M : M',$$

where M, M' denote the total masses of the two distributions.

7. Prove that, if two distributions of matter have the same external level surfaces, they have the same centre of gravity and the same principal axes of inertia at the centre of gravity.

[In Ex. 6 put $\phi = x$, and $\phi = yz$.]

8. Prove that, if V is the potential of a given distribution of matter, S a surface surrounding it, r the distance of an internal point from an element dS, ∂n an element of the outward drawn normal and θ the angle between r and ∂n, then

$$\int \frac{1}{r} \frac{\partial V}{\partial n} \, dS = \int \frac{V \cos \theta}{r^2} \, dS,$$

where the integration is over the whole surface.

9. Shew that, if V is a function which with its first derivative is finite and continuous throughout a region bounded by a surface S and r denotes distance from a point P within this region, then

$$4\pi V_P = \int V \frac{\partial}{\partial n}\left(\frac{1}{r}\right) dS - \int \frac{1}{r}\frac{\partial V}{\partial n} dS - \int \frac{1}{r} \nabla^2 V dv,$$

where ∂n denotes the outward drawn normal and the integrations are over the surface and through the volume.

10. Two gravitating solids M and M' have potentials V and V'; the inward normal force across any element dS of the surface of M is the sum of N and R due separately to M and M'; and the inward normal force across any element dS' of the surface of M' is the sum of N' and R', due separately to M and M'. Shew that the potential energy of their relative position differs by a constant from either

$$\frac{1}{4\pi}\int (RV - NV')\, dS \quad \text{or} \quad \frac{1}{4\pi}\int (R'V' - N'V)\, dS'. \quad \text{[M. T. 1897]}$$

11. If $\nabla^2 V_1 = 0$, $\nabla^2 V_0 = 0$ and $V_1 = V_0 = \overline{V}$ at a closed surface, while V_0, V_1 and their differential coefficients are everywhere finite without and within the surface respectively, and V_0 vanishes at infinity, find the potential in all space of the surface distribution σ where

$$4\pi\sigma = \frac{\partial V_1}{\partial n} - \frac{\partial V_0}{\partial n},$$

and ∂n is outwards.

Shew that $\int \overline{V} \frac{\partial}{\partial n}\left(\frac{1}{r}\right) dS$ is equal to $-\int \frac{1}{r}\frac{\partial V_1}{\partial n} dS$ when r is measured from an external point, and to $\int \frac{1}{r}\frac{\partial V_0}{\partial n} dS$ when r is measured from an internal point. [C. 1908]

---

12. Prove that, if ϕ denotes any function which with its first derivative is continuous, its mean value over a sphere of radius a exceeds its value at the centre of the sphere by

$$\frac{1}{4\pi}\int \left(\frac{1}{r} - \frac{1}{a}\right) \nabla^2\phi\, dv$$

integrated through the volume of the sphere, where r denotes distance from the centre. [Minchin.]

13. Assuming Gauss's theorem and Green's theorem and that a distribution of matter consisting of volume density ρ, surface density σ and separate particles has a potential function V which is infinite at each particle and satisfies the relations $\nabla^2 V = -4\pi\rho$ wherever there is volume density, and $\left(\dfrac{\partial V}{\partial n}\right)_1 - \left(\dfrac{\partial V}{\partial n}\right)_2 = 4\pi\sigma$ across each surface layer and is otherwise continuous, prove that the potential must be of the form

$$V = \Sigma \frac{m}{r} + \int \frac{\sigma \, dS}{r} + \int \frac{\rho \, dv}{r},$$

where m denotes the mass of a particle.

14. Prove the theorem of the equivalent layer [5·4] by applying Green's theorem to the region bounded externally by the infinite sphere and internally by the equipotential surface S and a small sphere round the point Q, and excluding the singularities of the potential function.

15. Prove that, if M be the mass of a given distribution of matter, U the potential of a closed equipotential surface entirely surrounding it, and E the exhaustion of potential energy in collecting the matter from a state of diffusion at infinity into its actual distribution, then $\frac{1}{2}MU < E$. [M. T. 1895]

16. A function V vanishes at all points outside a surface S. Shew that a distribution of matter $\rho = -\frac{1}{4\pi} \nabla^2 V$ within S and $\sigma = \frac{1}{4\pi} \frac{\partial V}{\partial n}$ upon S will have the function V for its potential. Here $\partial V/\partial n$ means the rate of increase of V in the direction approaching the surface from the inside. [C. 1931]

17. Prove that, if V_0 is the potential at points outside a surface S due to masses within the surface, and V_1 is the potential at points inside the surface due to masses outside it, the expression

$$\frac{1}{4\pi} \int (V_0 - V_1) \frac{\partial}{\partial n}\left(\frac{1}{r}\right) dS - \frac{1}{4\pi} \int \left(\frac{\partial V_0}{\partial n} - \frac{\partial V_1}{\partial n}\right) \frac{1}{r} \, dS$$

has the value V_0 at external points and the value V_1 at internal points, where r denotes distance from the point and ∂n an element of the outward normal to dS. [M. T. 1910]

18. V is the potential and R the attraction at any point due to a given distribution of matter none of which lies between two level surfaces $V = V_1$, $V = V_2$. The space between these surfaces is filled with matter of density $R^2 f(V)$. Prove that its potential at any external point P is $-4\pi U \int_{V_1}^{V_2} f(V) \, dV$, where U denotes the potential at P due to that part of the original distribution which lies inside the inner of the two level surfaces.

19. If V is the gravitational potential of a distribution of matter inside a closed surface S, shew that the value of V at a point P outside S is given in terms of its values on S by the formula

$$V_P = \frac{1}{4\pi} \int V \frac{\partial G}{\partial n} \, dS,$$

where the differentiation is along the outward normal to S, G vanishes on S, and, if r is distance measured from P, $G - 1/r$ is a harmonic function on and outside S. Give a physical interpretation of G and $\partial G/\partial n$.

Shew that, if V is the potential of a finite distribution of matter on one side of an infinite plane, and P is a point on the other side,

$$V_P = \frac{1}{2\pi} \int V_1 \, d\Omega,$$

where $d\Omega$ is the solid angle subtended at P by an element of the plane at which the value of V is V_1.   [M. T. 1934]

20. Use 5·61, 5·62 to shew that, if V be a function whose value is prescribed over a sphere S of radius a, and P a point inside the sphere at a distance f from the centre, then a solution of Laplace's equation inside S is given by

$$V_P = -\frac{a^2 - f^2}{4\pi a} \int \frac{V}{r^3} \, dS,$$

where r denotes the distance of dS from P.

21. If ϕ_1 and ϕ_2 are any scalar functions whose derivatives exist and are continuous at all points of a volume V and its boundary surface S, and if

$$I_{jk} = \int_V (\operatorname{grad} \phi_j \cdot \operatorname{grad} \phi_k) \, dV, \quad (j, k = 1, 2),$$

prove that $I_{11} I_{22} \geqslant I_{12}{}^2.$

If ϕ_1 is a given harmonic function in V and if $\phi_2 = \phi_1$ on S, shew that

$$I_{11} = I_{12}.$$

Hence shew that for such a ϕ_2 the necessary and sufficient condition that I_{22} should be a minimum is that ϕ_2 should be harmonic in V.

What is the physical interpretation of this result?   [M. T. 1935]

Chapter VI

HARMONIC FUNCTIONS

6·1. Any solution of Laplace's equation

$$\frac{\partial^2 V}{\partial x^2} + \frac{\partial^2 V}{\partial y^2} + \frac{\partial^2 V}{\partial z^2} = 0$$

which is homogeneous in x, y, z is called a **harmonic function** or a **spherical harmonic**. The degree of homogeneity is called the degree of the function. We shall be concerned with the case in which the degree is an integer.

The study of harmonic functions is an important branch of mathematical analysis. We shall only concern ourselves with so much of the theory as is useful for our immediate purpose and be content with arguments based upon physical ideas rather than attempt a rigorous treatment of the subject.

6·11. *If V is a harmonic function of degree n, then*

$$\frac{\partial^p}{\partial x^p} \frac{\partial^q}{\partial y^q} \frac{\partial^t}{\partial z^t} V$$

is a harmonic function of degree $n - p - q - t$.

For if we differentiate the equation $\nabla^2 V = 0$ p times with regard to x, q times with regard to y and t times with regard to z, we get

$$\nabla^2 \left\{ \frac{\partial^p}{\partial x^p} \frac{\partial^q}{\partial y^q} \frac{\partial^t}{\partial z^t} V \right\} = 0.$$

6·12. Surface and solid harmonics. In polar co-ordinates Laplace's equation may be written

$$\frac{\partial}{\partial r}\left(r^2 \frac{\partial V}{\partial r} \right) + \frac{1}{\sin\theta} \frac{\partial}{\partial \theta}\left(\sin\theta \frac{\partial V}{\partial \theta} \right) + \frac{1}{\sin^2\theta} \frac{\partial^2 V}{\partial \phi^2} = 0 \quad [\mathbf{1\cdot 68}\ (1)]$$

$$\ldots\ldots(1).$$

If we substitute $V = r^n S_n$, where S_n is independent of r, we get $\dfrac{\partial}{\partial r}\left(r^2 \dfrac{\partial V}{\partial r}\right) = n(n+1) r^n S_n$, so that (1) reduces to

$$\frac{\partial^2 S_n}{\partial \theta^2} + \cot \theta \frac{\partial S_n}{\partial \theta} + \frac{1}{\sin^2 \theta} \frac{\partial^2 S_n}{\partial \phi^2} + n(n+1) S_n = 0 \;\ldots(2),$$

or, if $\cos \theta = \mu$,

$$\frac{\partial}{\partial \mu}\left\{(1-\mu^2)\frac{\partial S_n}{\partial \mu}\right\} + \frac{1}{1-\mu^2}\frac{\partial^2 S_n}{\partial \phi^2} + n(n+1) S_n = 0 \;\ldots(3).$$

A solution S_n of equation (2) is called a **Laplace's function** or a **surface harmonic** of order n. Since $n(n+1)$ remains unchanged when we write $-(n+1)$ for n, there are two solutions of (1) of which S_n is a factor, namely $r^n S_n$ and $r^{-n-1} S_n$, and these are called **solid harmonics** of degree n and $-(n+1)$ respectively.

6·13. It follows that if U is a harmonic function of degree n, then U/r^{2n+1} is also a harmonic function. For we may write $U = r^n S_n$, so that $\dfrac{U}{r^{2n+1}} = \dfrac{S_n}{r^{n+1}}$, which is a harmonic function.

For example, it is obvious that xyz is a solution of Laplace's equation of the third degree; therefore xyz/r^7 is also a solution.

Similarly, if U is a harmonic function of degree $-(n+1)$, then $r^{2n+1}U$ is also a harmonic function. For we may write $U = r^{-n-1} S_n$, so that $r^{2n+1}U = r^n S_n$, which is harmonic.

These theorems combined with **6·11** make it easy to build up a great variety of harmonic functions.

Thus x^0 or 1 and θ or $\tan^{-1}(y/x)$ are both harmonic functions of degree zero. If we divide them by r, we get $\dfrac{1}{r}$ and $\dfrac{1}{r}\tan^{-1}\dfrac{y}{x}$ as harmonics of degree -1. Differentiating these with regard to x, y or z, we find that $\dfrac{x}{r^3}, \dfrac{y}{r^3}, \dfrac{z}{r^3}$ and $\dfrac{z}{r^3}\tan^{-1}\dfrac{y}{x}$ are harmonics of degree -2, and so on.

Again, if ∂h denotes an element of length whose direction cosines are l, m, n, so that $\dfrac{\partial}{\partial h} = l\dfrac{\partial}{\partial x} + m\dfrac{\partial}{\partial y} + n\dfrac{\partial}{\partial z}$, it follows that

$\dfrac{\partial}{\partial h}\left(\dfrac{1}{r}\right)$ is a harmonic function and h may be called its axis. Similarly, if $h_1,\ h_2,\ h_3,\ \ldots$ are any number of arbitrary directions, we can form a harmonic function $\dfrac{\partial}{\partial h_1}\dfrac{\partial}{\partial h_2}\dfrac{\partial}{\partial h_3}\cdots\left(\dfrac{1}{r}\right)$.

6·2. Legendre functions.* Let M be a fixed point (f,g,h) and P a variable point (x,y,z), then if $MP=R$, we know that the inverse distance $1/R$ is a solution of Laplace's equation, for it is the potential at P due to a particle of unit mass at M. Suppose that M lies on a sphere of radius a with its centre at the origin O, so that $f^2+g^2+h^2=a^2$, and let $OP=r$. Then we have

$$V=\frac{1}{R}=\frac{1}{\{(x-f)^2+(y-g)^2+(z-h)^2\}^{\frac{1}{2}}}\ \ldots\ldots\ldots(1).$$

Since $x^2+y^2+z^2=r^2$, the expression for V can be expanded in powers of $1/r$, by the extension of Taylor's theorem to functions of several variables, thus

$$V=\frac{1}{r}-\left(f\frac{\partial}{\partial x}+g\frac{\partial}{\partial y}+h\frac{\partial}{\partial z}\right)\frac{1}{r}+\ldots$$
$$+\frac{(-1)^n}{n!}\left(f\frac{\partial}{\partial x}+g\frac{\partial}{\partial y}+h\frac{\partial}{\partial z}\right)^n\frac{1}{r}+\ldots\ \ \ldots\ldots(2),$$

the series converging if $r>a$, i.e. if the point P is outside the sphere. But, if we write μ for the cosine of the angle MOP, we have

$$V=\frac{1}{(r^2-2ar\mu+a^2)^{\frac{1}{2}}},$$

and this can be expanded in powers of r/a when $r<a$, and in powers of a/r when $r>a$. Thus if we denote the coefficient of r^n/a^{n+1} or of a^n/r^{n+1} by P_n, we have

$$V=\frac{1}{a}\left\{1+\frac{r}{a}P_1+\ldots+\frac{r^n}{a^n}P_n+\ldots\right\}\quad r<a\quad\ldots\ldots(3)$$

and
$$\qquad V=\frac{1}{r}\left\{1+\frac{a}{r}P_1+\ldots+\frac{a^n}{r^n}P_n+\ldots\right\}\quad r>a\quad\ldots\ldots(4).$$

* Adrien Marie Legendre (1752–1833). French mathematician of great distinction.

The P's are clearly polynomials in μ, for they are the coefficients of successive powers of h in the expansion of

$$(1 - 2\mu h + h^2)^{-\frac{1}{2}} \quad \text{or}$$

$$1 + \tfrac{1}{2}(2\mu h - h^2) + \ldots + \frac{1.3\ldots(2n-1)}{2.4\ldots 2n}(2\mu h - h^2)^n + \ldots \quad \ldots(5).$$

By picking out coefficients of successive powers of h we find that

$$P_1 = \mu, \quad P_2 = \tfrac{1}{2}(3\mu^2 - 1), \quad P_3 = \tfrac{1}{2}(5\mu^3 - 3\mu), \ldots$$

and

$$P_n = \frac{2n!}{2^n n! \, n!}\left\{\mu^n - \frac{n(n-1)}{2(2n-1)}\mu^{n-2}\right.$$

$$\left. + \frac{n(n-1)(n-2)(n-3)}{2.4(2n-1)(2n-3)}\mu^{n-4} - \ldots\right\} \quad \ldots\ldots(6),$$

the series ending with a term in μ or a constant according as n is odd or even. For many purposes it is sufficient to know that P_n is a polynomial in μ of the form

$$P_n = A_n \mu^n + A_{n-2}\mu^{n-2} + \ldots \quad \ldots\ldots\ldots(7),$$

where A_n, A_{n-2}, ... are constants.

The function P_n is called the **Legendre function** or **Legendre coefficient of order n**. It is written $P_n(\mu)$ when it is desired to specify the variable. We observe that $\mu = \cos\theta$, where θ is an angle measured from an axis OM, called the *axis of the function*.

In particular, for $\theta = 0$ or π we have $\mu = \pm 1$, and

$$\left.\begin{array}{l} P_n(1) \quad = \text{coefficient of } h^n \text{ in } \dfrac{1}{1-h} = 1, \\[2mm] P_n(-1) = \ldots\ldots\ldots\ldots\ldots\ldots\ldots \dfrac{1}{1+h} = (-1)^n \end{array}\right\} \quad \ldots(8).$$

If we write $\mu = \cos\theta$, then

$$(1 - 2\mu h + h^2)^{-\frac{1}{2}} = (1 - he^{i\theta})^{-\frac{1}{2}}(1 - he^{-i\theta})^{-\frac{1}{2}}$$

$$= \left(1 + \tfrac{1}{2}he^{i\theta} + \frac{1.3}{2.4}h^2 e^{2i\theta} + \ldots\right)\left(1 + \tfrac{1}{2}he^{-i\theta} + \frac{1.3}{2.4}h^2 e^{-2i\theta} + \ldots\right).$$

It follows that P_n is of the form

$$P_n = B_n \cos n\theta + B_{n-2} \cos (n-2)\,\theta + ...,$$

where B_n, B_{n-2}, ... are positive. Hence the numerical value of P_n does not exceed the number obtained by putting $\theta = 0$ in the above expression.

Thus $\qquad |P_n(\mu)| \leqslant P_n(1)$; i.e. $|P_n(\mu)| \leqslant 1$ ....(9).

6·21. $r^n P_n$ **and** $r^{-n-1}P_n$ **are harmonic functions.** Since $\mu = \cos MOP = (fx + gy + hz)/ar$, from **6·2** (7) we have

$$r^n P_n = \frac{A_n}{a^n}(fx + gy + hz)^n$$

$$+ \frac{A_{n-2}}{a^{n-2}}(fx + gy + hz)^{n-2}(x^2 + y^2 + z^2) + ... \quad ......(1),$$

expressing $r^n P_n$ as a homogeneous rational integral function of the co-ordinates of P.

Further, we observe that in **6·2**, (2) and (4) represent the same expansion of V, so that by comparing the terms in r^{-n-1} we have

$$\frac{a^n}{r^{n+1}}P_n = \frac{(-1)^n}{n!}\left(f\frac{\partial}{\partial x} + g\frac{\partial}{\partial y} + h\frac{\partial}{\partial z}\right)^n \frac{1}{r} \quad ......(2).$$

But $1/r$ is a harmonic function, therefore by **6·11** every term on the right-hand side of (2) is a harmonic function of degree $-(n+1)$. Hence $r^{-n-1}P_n$ is a harmonic function, and by **6·13** $r^n P_n$ is also a harmonic function.

6·22. Rodrigues's formula for P_n. * *To prove that*

$$P_n(\mu) = \frac{1}{2^n \cdot n!}\frac{d^n}{d\mu^n}\{(\mu^2 - 1)^n\}.$$

This formula may be verified by expanding $(\mu^2 - 1)^n$ by the binomial theorem and differentiating the result n times; or we may proceed as follows:

Let $\qquad\qquad 1 - hu = (1 - 2\mu h + h^2)^{\frac{1}{2}}$ ....(1),

so that $\qquad\qquad u = \mu + \tfrac{1}{2}h(u^2 - 1)$.

* *Corresp. sur l'École polytechnique*, vol. III (1814–16), pp. 361–85.

By a theorem of Lagrange* u can be expanded in powers of h in the form

$$u = \mu + \tfrac{1}{2}h\,(\mu^2 - 1) + \ldots + \frac{h^n}{2^n n!}\left(\frac{d}{d\mu}\right)^{n-1}(\mu^2 - 1)^n + \ldots \quad \ldots(2).$$

But by differentiating (1) we have

$$\frac{du}{d\mu} = (1 - 2\mu h + h^2)^{-\frac{3}{2}} = \Sigma h^n P_n \quad \ldots\ldots\ldots\ldots\ldots(3).$$

Then by differentiating (2) and comparing the coefficients of h^n in (3) and in the result of differentiating (2), we obtain the required formula.

6·23. The equation $P_n(\mu) = 0$. Since the equation $(\mu^2 - 1)^n = 0$ has n roots equal to 1 and n roots equal to -1, the equation $\dfrac{d}{d\mu}(\mu^2 - 1)^n = 0$ has $n-1$ roots equal to -1, $n-1$ roots equal to 1, and a root ξ between -1 and 1. Then $\dfrac{d^2}{d\mu^2}(\mu^2 - 1)^n = 0$ has $n-2$ roots equal to -1, $n-2$ roots equal to 1, a root between -1 and ξ and a root between ξ and 1. Continuing the process it follows that the n roots of the equation $P_n(\mu) = 0$ are all real and lie between ± 1.

6·24. Differential equation for P_n. Since $r^n P_n$ is a harmonic function and P_n is a function of θ only, $r^n P_n$ satisfies Laplace's equation in the form

$$\frac{\partial}{\partial r}\left(r^2 \frac{\partial V}{\partial r}\right) + \frac{1}{\sin\theta}\frac{\partial}{\partial\theta}\left(\sin\theta \frac{\partial V}{\partial\theta}\right) = 0,$$

and as in **6·12** we have

$$\frac{1}{\sin\theta}\frac{\partial}{\partial\theta}\left(\sin\theta \frac{\partial P_n}{\partial\theta}\right) + n\,(n+1)\,P_n = 0,$$

or putting $\cos\theta = \mu$,

$$\frac{d}{d\mu}\left\{(1 - \mu^2)\frac{dP_n}{d\mu}\right\} + n\,(n+1)\,P_n = 0 \quad \ldots\ldots\ldots(1).$$

This is commonly known as *Legendre's equation*. It can be solved in series in the form **6·2** (6).

6·25. Integral theorems. *To prove that*

$$\int_{-1}^{1} P_n(\mu)\,P_m(\mu)\,d\mu = 0, \quad n \neq m,$$

and

$$\int_{-1}^{1} P_n{}^2(\mu)\,d\mu = \frac{2}{2n+1}.$$

* Whittaker and Watson, *Modern Analysis*, **7·32**.

From 6·24 we have

$$\frac{d}{d\mu}\left\{(1-\mu^2)\frac{dP_n}{d\mu}\right\}+n(n+1)P_n=0$$

and

$$\frac{d}{d\mu}\left\{(1-\mu^2)\frac{dP_m}{d\mu}\right\}+m(m+1)P_m=0.$$

Therefore

$$\{n(n+1)-m(m+1)\}\int_{-1}^{1}P_nP_m\,d\mu$$

$$=\int_{-1}^{1}\left[P_n\frac{d}{d\mu}\left\{(1-\mu^2)\frac{dP_m}{d\mu}\right\}-P_m\frac{d}{d\mu}\left\{(1-\mu^2)\frac{dP_n}{d\mu}\right\}\right]d\mu,$$

and, integrating by parts, this

$$=\left[(1-\mu^2)\left(P_n\frac{dP_m}{d\mu}-P_m\frac{dP_n}{d\mu}\right)\right]_{-1}^{1}$$

$$-\int_{-1}^{1}(1-\mu^2)\left(\frac{dP_n}{d\mu}\frac{dP_m}{d\mu}-\frac{dP_m}{d\mu}\frac{dP_n}{d\mu}\right)d\mu=0.$$

Again, to find $\int_{-1}^{1}P_n{}^2(\mu)\,d\mu$, we have

$$\left\{\sum_0^\infty h^nP_n(\mu)\right\}^2=(1-2\mu h+h^2)^{-1}.$$

Integrate between limits -1 to 1 and use the fact that the integrals of all products P_nP_m $(n\neq m)$ vanish, and we get

$$\int_{-1}^{1}P_n{}^2(\mu)\,d\mu=\text{coeff. of }h^{2n}\text{ in }\int_{-1}^{1}\frac{d\mu}{1-2\mu h+h^2}$$

$$=\dots\dots\dots\dots -\frac{1}{h}\log\frac{1-h}{1+h}$$

$$=\frac{2}{2n+1}.$$

6·26. From the expression for $P_n(\mu)$ in powers of μ, it is clear that by solving a set of linear equations any positive integral power of μ can be expressed in terms of a finite number of P's in the form

$$\mu^n=a_0P_n+a_2P_{n-2}+a_4P_{n-4}+\dots$$

and consequently, from 6·25, if $n<m$,

$$\int_{-1}^{1}\mu^nP_m^\cdot(\mu)\,d\mu=0.$$

And similarly, if $F(\mu)$ is a polynomial of degree less than m,

$$\int_{-1}^{1} F(\mu) P_m(\mu) \, d\mu = 0.$$

6·261. If a polynomial in μ is expressed in terms of Legendre coefficients, say in the form

$$F(\mu) = \Sigma A_n P_n(\mu),$$

the coefficient A_n can be found by multiplying both sides by $P_n(\mu)$ and integrating between limits -1 and 1. For then by **6·25** all the terms on the right except the term in A_n vanish, and

$$A_n = \frac{2n+1}{2} \int_{-1}^{1} F(\mu) P_n(\mu) \, d\mu.$$

6·27. Recurrence formulae. These are a set of formulae connecting successive P's or their derivatives.

We begin with the formula

$$(1 - 2\mu h + h^2)^{-\frac{1}{2}} = 1 + \sum_{1}^{\infty} h^n P_n(\mu) \quad \dots\dots\dots\dots(1).$$

Differentiation with regard to h gives

$$(\mu - h)(1 - 2\mu h + h^2)^{-\frac{3}{2}} = \sum_{1}^{\infty} n h^{n-1} P_n(\mu) \quad \dots\dots\dots(2).$$

Multiplying by $(1 - 2\mu h + h^2)$, we get

$$(\mu - h)\left\{1 + \sum_{1}^{\infty} h^n P_n(\mu)\right\} = (1 - 2\mu h + h^2)\sum_{1}^{\infty} n h^{n-1} P_n(\mu),$$

then equating the coefficients of h^n gives

$$(n+1) P_{n+1} - (2n+1) \mu P_n + n P_{n-1} = 0 \dots\dots\dots[A].$$

Again, differentiate (1) with regard to μ and we have

$$h(1 - 2\mu h + h^2)^{-\frac{3}{2}} = \sum_{1}^{\infty} h^n \frac{dP_n}{d\mu},$$

and therefore, from (2),

$$h \sum_{1}^{\infty} n h^{n-1} P_n = (\mu - h) \sum_{1}^{\infty} h^n \frac{dP_n}{d\mu}.$$

Whence, by equating coefficients of h^n

$$n P_n = \mu \frac{dP_n}{d\mu} - \frac{dP_{n-1}}{d\mu} \quad \dots\dots\dots\dots\dots[B].$$

Again, by differentiating [A] we get

$$(n+1)\frac{dP_{n+1}}{d\mu} - (2n+1) P_n - (2n+1)\mu\frac{dP_n}{d\mu} + n\frac{dP_{n-1}}{d\mu} = 0,$$

and eliminating $dP_n/d\mu$ from this result and [B] gives

$$(2n+1)P_n = \frac{dP_{n+1}}{d\mu} - \frac{dP_{n-1}}{d\mu} \quad \dots\dots\dots\dots[\text{C}].$$

Subtracting [B] from [C] gives

$$(n+1)P_n = \frac{dP_{n+1}}{d\mu} - \mu\frac{dP_n}{d\mu} \quad \dots\dots\dots\dots[\text{D}].$$

In [D] change $n+1$ into n and subtract μ times [B] and we get

$$(\mu^2-1)\frac{dP_n}{d\mu} = n\mu P_n - nP_{n-1} \quad \dots\dots\dots\dots[\text{E}].$$

Integrating [C] gives

$$\int P_n(\mu)\,d\mu = \frac{P_{n+1}-P_{n-1}}{2n+1} + \text{const.} \quad \dots\dots\dots\dots[\text{F}].$$

Lastly, by adding successive formulae of the form [C] we get

$$\frac{dP_n}{d\mu} = (2n-1)P_{n-1} + (2n-5)P_{n-3} + (2n-9)P_{n-5} + \dots,$$

the series ending with $3P_1$ if n is even, and with P_0 or 1 if n is odd
$$\dots\dots[\text{G}].$$

6·271. Example. *Use the theorems of* 6·25 *to prove* 6·27 [G].

Since P_n is a polynomial of the nth degree in μ, therefore $dP_n/d\mu$ is a polynomial of the $(n-1)$th degree and by solving linear equations is expressible in the form

$$\frac{dP_n}{d\mu} = a_{n-1}P_{n-1} + a_{n-3}P_{n-3} + \dots.$$

To find any coefficient, say a_m, multiply by P_m and integrate between limits -1 and 1; then

$$\int_{-1}^{1} P_m\frac{dP_n}{d\mu}\,d\mu = a_m\int_{-1}^{1} P_m^2\,d\mu,$$

since by 6·25 every other term vanishes,

$$= \frac{2a_m}{2m+1}, \quad \text{by } 6\cdot25 \quad \dots\dots\dots\dots\dots(1).$$

But since $m < n$, $dP_m/d\mu$ is of the form $c_{m-1}P_{m-1} + c_{m-3}P_{m-3} + \dots$, where the suffixes are all less than n. Therefore

$$\int_{-1}^{1} P_n\frac{dP_m}{d\mu}\,d\mu = 0 \quad \dots\dots\dots\dots\dots(2).$$

By adding (1) and (2) we get

$$\frac{2a_m}{2m+1} = \int_{-1}^{1}\left(P_m\frac{dP_n}{d\mu} + P_n\frac{dP_m}{d\mu}\right)d\mu$$

$$= \left[P_m P_n\right]_{-1}^{1} = 2,$$

since $n-m$ is odd, 6·2 (8).

Therefore $a_m = 2m + 1$, and

$$\frac{dP_n}{d\mu} = (2n-1) P_{n-1} + (2n-5) P_{n-3} + \dots.$$

6·28. Definite integral forms for P_n. There are two expressions for P_n as a definite integral, due to Laplace,* which may be obtained as follows: we have

$$\int_0^\pi \frac{d\phi}{a + b \cos \phi} = \frac{\pi}{\sqrt{(a^2 - b^2)}},$$

so that, by putting $a = 1 - \mu h$ and $b = \pm h \sqrt{(\mu^2 - 1)}$, we get

$$\frac{1}{\sqrt{(1 - 2\mu h + h^2)}} = \frac{1}{\sqrt{\{(1 - \mu h)^2 - h^2 (\mu^2 - 1)\}}}$$

$$= \frac{1}{\pi} \int_0^\pi \frac{d\phi}{1 - \mu h \pm h \sqrt{(\mu^2 - 1)} \cos \phi}.$$

Expanding both expressions in powers of h, we get

$$\sum_0^\infty h^n P_n = \frac{1}{\pi} \int_0^\pi \sum_0^\infty h^n \{\mu \mp \sqrt{(\mu^2 - 1)} \cos \phi\}^n \, d\phi,$$

and, by comparing the coefficients of h^n,

$$P_n = \frac{1}{\pi} \int_0^\pi \{\mu \mp \sqrt{(\mu^2 - 1)} \cos \phi\}^n \, d\phi \quad \dots\dots(1).$$

$\sqrt{(\mu^2 - 1)}$ is imaginary for values of μ between -1 and 1, but if the binomial expression is expanded the integrals of all odd powers of $\cos \phi$ vanish and a real result is obtained.

By writing

$$\cos \phi = \{\mu \cos \psi \pm \sqrt{(\mu^2 - 1)}\} / \{\mu \pm \sqrt{(\mu^2 - 1)} \cos \psi\},$$

the result may be transformed to

$$P_n = \frac{1}{\pi} \int_0^\pi \{\mu \pm \sqrt{(\mu^2 - 1)} \cos \psi\}^{-n-1} \, d\psi \quad \dots\dots(2).$$

6·29. After this lengthy discussion of the properties of the Legendre polynomials we revert to their introduction in **6·2** as a special form of surface harmonic, namely a solution of

* *Mécanique Céleste*, liv. XI, ch. II, p. 33 [1823].

6·12 (2) independent of ϕ. But we must point out that, if in **6·2** we take M as the point (a, θ', ϕ') and P as (r, θ, ϕ), then

$$V = \{r^2 + a^2 - 2ar\,[\cos\theta\cos\theta' + \sin\theta\sin\theta'\cos(\phi - \phi')]\}^{-\frac{1}{2}}$$
$$\dots\dots(1)$$

$$= (r^2 + a^2 - 2ar\lambda)^{-\frac{1}{2}} \quad\dots\dots\dots\dots\dots\dots(2),$$

where
$$\left.\begin{array}{l} \lambda = \mu\mu' + \sqrt{(1-\mu^2)}\,\sqrt{(1-\mu'^2)}\cos(\phi - \phi'), \\ \mu = \cos\theta, \quad \mu' = \cos\theta' \end{array}\right\} \quad\dots(3).$$

This as before is expansible in powers of r/a or a/r, and it may be shewn, as in **6·22**, that the coefficient of r^n/a^{n+1} or of a^n/r^{n+1} is $\dfrac{1}{2^n n!}\dfrac{d^n}{d\lambda^n}\{(\lambda^2 - 1)^n\}$, where λ is given by (3).

Expressed in this way the Legendre function $P_n(\lambda)$ is a symmetrical function of θ, ϕ and θ', ϕ', i.e. of the angular co-ordinates of P and M, and it is only when the polar axis of co-ordinates coincides with OM or OP that we have $\lambda = \mu$ or $\lambda = \mu'$ and the Legendre function takes the form of a polynomial independent of ϕ and ϕ'.

6·3. Applications of the Legendre functions.

Since the Legendre function $P_n(\mu)$ is a function of μ or $\cos\theta$ only and $r^n P_n$ and $r^{-n-1} P_n$ are solutions of Laplace's equation, it is evident that these functions may prove to be of use in connection with the potentials of bodies with an axis of symmetry. In this connection there is a **theorem of Legendre**: *If the potential of a body with an axis of symmetry is known at points on a finite length of the axis outside the body it can be calculated in terms of the Legendre functions at all points that can be reached from this length of the axis without traversing matter.* It is to be understood that there is symmetry both as regards the density and the shape of the body.

Taking the axis of symmetry as the axis Oz, let Q be a point on the axis outside the body at a distance z from the origin. And suppose that the potential at Q is expressible in a series in the form

$$U = a_0 + \frac{b_0}{z} + a_1 z + \frac{b_1}{z^2} + \dots + a_n z^n + \frac{b_n}{z^{n+1}} + \dots \quad\dots(1)$$

and that this series converges for the z's pertaining to a certain length of the axis.

Also let V denote the potential at a point P, where $OP = r$ and $QOP = \theta$; and let

$$V' = a_0 + \frac{b_0}{r} + \left(a_1 r + \frac{b_1}{r^2}\right) P_1 + \dots + \left(a_n r^n + \frac{b_n}{r^{n+1}}\right) P_n + \dots$$
$$\dots\dots(2).$$

Then V' is a harmonic function [6·21], so that $V - V'$ is also a harmonic function. Now let P move up to Q, then V becomes U and V' becomes the series on the right of (1) (since at Q $r = z$ and $\theta = 0$), so that at Q, and other points on Oz for which (1) converges, $V - V'$ vanishes. Therefore by 4·3 (v) the potential function $V - V'$ vanishes at all points that can be reached in passing from Q and other neighbouring points on the axis without passing through the attracting matter or through a region in which (1) ceases to converge. We conclude therefore that the potential at any such point P is given by

$$V = a_0 + \frac{b_0}{r} + \left(a_1 r + \frac{b_1}{r^2}\right) P_1 + \dots + \left(a_n r^n + \frac{b_n}{r^{n+1}}\right) P_n + \dots$$
$$\dots\dots(3),$$

and this series can clearly be written down, whenever (1) has been calculated, by writing r for z and introducing the appropriate P in each term.

It is clear that (3) is absolutely convergent for all values of r which make (1) absolutely convergent when r is written for z, since $|P_n| \leqslant 1$.

6·31. Potential of a circular ring.

The potential of a circular ring of radius a and line density m at a point on its axis at a distance z from its centre is $2\pi am/(a^2 + z^2)^{\frac{1}{2}}$. This can be expanded in powers of z/a or a/z according as z is less or greater than a, so that on the axis the potential is given by either

$$U_1 = 2\pi m \left(1 - \frac{1}{2}\frac{z^2}{a^2} + \frac{1 \cdot 3}{2 \cdot 4}\frac{z^4}{a^4} - \dots\right), \quad z < a,$$

or

$$U_2 = 2\pi m \left(\frac{a}{z} - \frac{1}{2}\frac{a^3}{z^3} + \frac{1 \cdot 3}{2 \cdot 4}\frac{a^5}{z^5} - \dots\right), \quad z > a.$$

Therefore the potential at any point (r, θ) is either

$$V_1 = 2\pi m \left(1 - \frac{1}{2}\frac{r^2}{a^2}P_2 + \frac{1\cdot3}{2\cdot4}\frac{r^4}{a^4}P_4 - \dots\right), \quad r < a,$$

or $\qquad V_2 = 2\pi m \left(\frac{a}{r} - \frac{1}{2}\frac{a^3}{r^3}P_2 + \frac{1\cdot3}{2\cdot4}\frac{a^5}{r^5}P_4 - \dots\right), \quad r > a.$

6·32. Potential of a circular disc. With the same notation the potential at a point on the positive axis of z is, by **2·51**, $2\pi m\{\sqrt{(z^2 + a^2)} - z\}$, where m is the mass of unit area. This can be expanded in the forms

$$U_1 = 2\pi m \left\{a - z + \frac{1}{2}\frac{z^2}{a} - \frac{1\cdot1}{2\cdot4}\frac{z^4}{a^3} + \frac{1\cdot1\cdot3}{2\cdot4\cdot6}\frac{z^6}{a^5} - \dots\right\}, \quad z < a,$$

and $\quad U_2 = 2\pi m \left\{\frac{1}{2}\frac{a^2}{z} - \frac{1\cdot1}{2\cdot4}\frac{a^4}{z^3} + \frac{1\cdot1\cdot3}{2\cdot4\cdot6}\frac{a^6}{z^5} - \dots\right\}, \quad z > a.$

Therefore the potential at (r, θ) is given by

$$V_1 = 2\pi m \left\{a - rP_1 + \frac{1}{2}\frac{r^2}{a}P_2 - \frac{1\cdot1}{2\cdot4}\frac{r^4}{a^3}P_4 + \frac{1\cdot1\cdot3}{2\cdot4\cdot6}\frac{r^6}{a^5}P_6 - \dots\right\}, \quad r < a$$

$$\dots\dots(1),$$

and

$$V_2 = 2\pi m \left\{\frac{1}{2}\frac{a^2}{r} - \frac{1\cdot1}{2\cdot4}\frac{a^4}{r^3}P_2 + \frac{1\cdot1\cdot3}{2\cdot4\cdot6}\frac{a^6}{r^5}P_4 - \dots\right\}, \quad r > a \quad \dots\dots(2).$$

If $r > a$, the expression (2) is valid for $0 \leqslant \theta \leqslant \pi$, since we can move from the point $(r, 0)$ to the point (r, θ) keeping $r > a$ without crossing the disc.

If $r < a$, and $\theta > \frac{1}{2}\pi$, we cannot move from $(r, 0)$ to (r, θ) keeping $r < a$, without crossing the disc. Thus (1) only holds if $0 \leqslant \theta \leqslant \frac{1}{2}\pi$.

If $\frac{1}{2}\pi < \theta \leqslant \pi$, then, by symmetry, $V_1(r, \theta) = V_1(r, \pi - \theta)$.

Now $P_1(\pi - \theta) = -P_1(\theta)$, and $P_2, P_4, \dots$ are unaltered if θ is replaced by $\pi - \theta$.

Hence when $r < a$ and $\frac{1}{2}\pi < \theta \leqslant \pi$

$$V_1 = 2\pi m \left\{a + rP_1 + \frac{1}{2}\frac{r^2}{a}P_2 - \frac{1\cdot1}{2\cdot4}\frac{r^4}{a^3}P_4 + \dots\right\}.$$

Thus, as in **2·51**, the form of the potential is discontinuous in crossing the disc.

6·33. Example. *Find, in the form of an infinite series, the potential due to a flat circular annulus (e.g. Saturn's ring) at points moderately near the centre of the ring.*

If a small but massive planet occupies the centre of the ring, shew that the maximum deflection of the plumb line on its surface (on account of the attraction of the ring) will occur in latitude 45°, approximately.

[M. T. 1916]

Let a, b be the inner and outer radii of the annulus. Its potential is the difference between the potentials of discs of radii a and b. Taking the axis of the annulus as axis of z with the centre as origin, the potential at the point $(0, 0, z)$ is

$$U = 2\pi m \{\sqrt{(b^2 + z^2)} - \sqrt{(a^2 + z^2)}\},$$

where m is the mass of unit area of the annulus.

For $z < a$, this gives

$$U = 2\pi mb \left(1 + \frac{1}{2}\frac{z^2}{b^2} - \dots\right) - 2\pi ma \left(1 + \frac{1}{2}\frac{z^2}{a^2} - \dots\right).$$

Therefore, at points near enough to the centre for higher powers of r/a to be neglected, the potential is

$$V = 2\pi mb \left(1 + \frac{1}{2}\frac{r^2}{b^2}P_2\right) - 2\pi ma \left(1 + \frac{1}{2}\frac{r^2}{a^2}P_2\right).$$

If the planet be of radius c, the deflecting force on the plumb line at a point on its surface at an angular distance θ from Oz is $\partial V/c\partial\theta$, and if c is small compared with a, this is

$$\frac{2\pi mb}{c}\frac{c^2}{2b^2}\frac{dP_2}{d\theta} - \frac{2\pi ma}{c}\cdot\frac{c^2}{2a^2}\frac{dP_2}{d\theta} = \pi mc\left(\frac{1}{b} - \frac{1}{a}\right)\frac{dP_2}{d\theta}.$$

But $P_2 = \frac{1}{2}(3\cos^2\theta - 1)$, so that the deflecting force is

$$-\tfrac{3}{2}\pi mc\left(\frac{1}{b} - \frac{1}{a}\right)\sin 2\theta,$$

and has its greatest value when $\theta = 45°$.

6·4. Harmonic functions in general. We now turn back to the subject of harmonic functions in general. From **6·1** H_n will denote a harmonic function of degree n if it is homogeneous of degree n in x, y, z and satisfies Laplace's equation. If it is a rational integral function of degree n in x, y, z, it contains $\frac{1}{2}(n+1)(n+2)$ coefficients; but since $\nabla^2 H_n = 0$, and this relation is of degree $n-2$, there are $\frac{1}{2}n(n-1)$ relations between the coefficients. Therefore $\frac{1}{2}(n+1)(n+2) - \frac{1}{2}n(n-1)$, or $2n+1$, of the coefficients are independent, so that there are $2n+1$ independent rational integral harmonic functions of degree n.

6·41. Integral theorem for surface harmonics. *To prove that* $\int S_m S_n \, d\omega = 0$, *where* S_m, S_n *are surface harmonics of different orders, $d\omega$ is an elementary solid angle, and the integral extends to the whole surface of the unit sphere.*

Apply Green's theorem **5·1** (1) to the region bounded by a sphere of radius a with its centre at the origin, putting $V = r^m S_m$ and $V' = r^n S_n$, so that V, V' are harmonic. It follows that

$$\int \left(V \frac{\partial V'}{\partial r} - V' \frac{\partial V}{\partial r} \right) dS = 0,$$

when the integration is over the sphere $r = a$, so that $dS = a^2 d\omega$.

The relation reduces to

$$(n - m) \, a^{m+n+1} \int S_m S_n \, d\omega = 0,$$

and, since $n \neq m$, therefore

$$\int S_m S_n \, d\omega = 0 \quad \dots \dots \dots \dots \dots (1).$$

Particular cases: (i) Since P_n is also a surface harmonic, therefore

$$\int S_m P_n \, d\omega = 0 \quad \dots \dots \dots \dots \dots (2).$$

(ii) Since $P_0 = 1$, therefore

$$\int S_m \, d\omega = 0, \quad m \neq 0 \quad \dots \dots \dots \dots (3).$$

(iii) Since P_m, P_n may be taken as special cases of S_m, S_n, and we may write $d\omega = \sin \theta \, d\theta \, d\phi = -d\mu \, d\phi$, the integral over the sphere in this case becomes $\int_{-1}^{1} \int_{0}^{2\pi} P_m P_n \, d\mu \, d\phi$, or $2\pi \int_{-1}^{1} P_m P_n \, d\mu$, so that (1) reduces in this case to the theorem of **6·25**, $\int_{-1}^{1} P_m P_n \, d\mu = 0$, $m \neq n$.

6·42. Tesseral harmonics. Reverting to the equation **6·12** (2) for surface harmonics, viz.

$$\frac{\partial^2 S_n}{\partial \theta^2} + \cot \theta \frac{\partial S_n}{\partial \theta} + \frac{1}{\sin^2 \theta} \frac{\partial^2 S_n}{\partial \phi^2} + n(n+1) S_n = 0 \quad \dots (1),$$

we proceed to examine solutions of the form $S_n = \Theta\Phi$, where Θ is a function of θ alone and Φ a function of ϕ alone. On multiplying the equation by $\sin^2\theta/\Theta\Phi$, it becomes

$$\frac{\sin^2\theta}{\Theta}\left\{\frac{\partial^2\Theta}{\partial\theta^2} + \cot\theta\frac{\partial\Theta}{\partial\theta} + n(n+1)\Theta\right\} + \frac{1}{\Phi}\frac{\partial^2\Phi}{\partial\phi^2} = 0 \quad\dots(2).$$

We must therefore have

$$\frac{\sin^2\theta}{\Theta}\left\{\frac{\partial^2\Theta}{\partial\theta^2} + \cot\theta\frac{\partial\Theta}{\partial\theta} + n(n+1)\Theta\right\} = c$$

and

$$\frac{1}{\Phi}\frac{\partial^2\Phi}{\partial\phi^2} = -c.$$

The latter equation has single-valued solutions when $c = p^2$, viz.

$$\Phi = C_p\cos p\phi + D_p\sin p\phi \quad\dots\dots\dots\dots\dots(3);$$

and Θ is then given in terms of θ by

$$\frac{d^2\Theta}{d\theta^2} + \cot\theta\frac{d\Theta}{d\theta} + \left\{n(n+1) - \frac{p^2}{\sin^2\theta}\right\}\Theta = 0,$$

or, in terms of μ, by

$$\frac{d}{d\mu}\left\{(1-\mu^2)\frac{d\Theta}{d\mu}\right\} + \left\{n(n+1) - \frac{p^2}{1-\mu^2}\right\}\Theta = 0 \quad\dots\dots(4).$$

For $p = 0$ this is Legendre's equation with a solution $P_n(\mu)$, where

$$\frac{d}{d\mu}\left\{(1-\mu^2)\frac{dP_n}{d\mu}\right\} + n(n+1)P_n = 0 \quad\dots\dots\dots\dots(5).$$

Differentiate (5) p times with regard to μ and multiply the result by $(1-\mu^2)^{\frac{1}{2}p}$. This gives

$$(1-\mu^2)^{\frac{1}{2}p+1}\frac{d^{p+2}P_n}{d\mu^{p+2}} - 2(p+1)\mu(1-\mu^2)^{\frac{1}{2}p}\frac{d^{p+1}P_n}{d\mu^{p+1}}$$

$$+ \{n(n+1) - p(p+1)\}(1-\mu^2)^{\frac{1}{2}p}\frac{d^p P_n}{d\mu^p} = 0 \quad\dots\dots(6).$$

Let

$$z = (1-\mu^2)^{\frac{1}{2}p}\frac{d^p P_n}{d\mu^p};$$

then

$$\frac{dz}{d\mu} = (1-\mu^2)^{\frac{1}{2}p}\frac{d^{p+1}P_n}{d\mu^{p+1}} - p\mu(1-\mu^2)^{\frac{1}{2}p-1}\frac{d^p P_n}{d\mu^p}.$$

Therefore

$$\frac{d}{d\mu}\left\{(1-\mu^2)\frac{dz}{d\mu}\right\}$$

$$= (1-\mu^2)^{\frac{1}{2}p+1}\frac{d^{p+2}P_n}{d\mu^{p+2}} - 2(p+1)\mu(1-\mu^2)^{\frac{1}{2}p}\frac{d^{p+1}P_n}{d\mu^{p+1}}$$

$$- p(1-\mu^2)^{\frac{1}{2}p}\frac{d^pP_n}{d\mu^p} + p^2\mu^2(1-\mu^2)^{\frac{1}{2}p-1}\frac{d^pP_n}{d\mu^p} \quad ...(7).$$

By subtracting (6) from the right-hand side of (7), we get

$$\frac{d}{d\mu}\left\{(1-\mu^2)\frac{dz}{d\mu}\right\}$$

$$= \left\{-n(n+1)+p(p+1)-p+\frac{p^2\mu^2}{1-\mu^2}\right\}(1-\mu^2)^{\frac{1}{2}p}\frac{d^pP_n}{d\mu^p}$$

$$= -\left\{n(n+1)-\frac{p^2}{1-\mu^2}\right\}z,$$

so that z is a solution of equation (4). It is convenient to denote solutions of this form by $X_n{}^p$ and, using Rodrigues's formula for P_n, write

$$X_n{}^p = \frac{(1-\mu^2)^{\frac{1}{2}p}}{2^n n!\, p!}\frac{d^{n+p}}{d\mu^{n+p}}(\mu^2-1)^n, \quad p=0, 1, \ldots n \; ...(8).$$

It is clear that $X_n{}^p$ is a rational integral function of $\sin\theta$ and $\cos\theta$, and that it vanishes if $n+p>2n$, i.e. if $p>n$. Also since $\Theta\Phi$ is a surface harmonic, therefore

$$X_n{}^0, \; X_n{}^1{}^{\cos}_{\sin}\phi, \; \ldots \; X_n{}^p{}^{\cos}_{\sin}p\phi, \; \ldots \; X_n{}^n{}^{\cos}_{\sin}n\phi$$

are $2n+1$ surface harmonics. They are called *Tesseral harmonics*. That they are linearly independent follows from the fact that if we had a relation

$$A_0X_n{}^0 + A_1X_n{}^1\cos\phi + \ldots + A_nX_n{}^n\cos n\phi$$

$$+ B_1X_n{}^1\sin\phi + \ldots + B_nX_n{}^n\sin n\phi = 0,$$

it would be sufficient to multiply by ${}^{\cos}_{\sin}p\phi$ and integrate from 0 to 2π, to shew that $A_p=0$, $B_p=0$ for $p=1, 2, \ldots n$; and then $A_0=0$.

6·43. H_n **expressed in tesseral harmonics.** From **6·4**, every surface harmonic of order n can be expressed as a linear combination of $2n+1$ linearly independent surface harmonics, so that

$$S_n = A_0 X_n{}^0 + A_1 X_n{}^1 \cos\phi + \ldots + A_n X_n{}^n \cos n\phi$$
$$+ B_1 X_n{}^1 \sin\phi + \ldots + B_n X_n{}^n \sin n\phi \quad \ldots\ldots(1),$$

where the A's and B's are constants.

Also since $H_n = r^n S_n$, therefore H_n is expressible in terms of harmonics of the form $r^n X_n{}^p \, {\cos \atop \sin} \, p\phi$, and it is easy to see that these are polynomials in x, y, z.

Thus **6·42** (8) may be written

$$X_n{}^p = \frac{\sin^p\theta}{p!}\frac{d^p P_n}{d\mu^p} \quad\ldots\ldots\ldots\ldots\ldots\ldots(2),$$

and taking for P_n its expansion in powers of μ [**6·2** (6)] we see that $X_n{}^p$ is the sum of terms of the form $\sin^p\theta\,(\cos\theta)^{n-p-2k}$; and

$$r^n \sin p\phi \sin^p\theta\,(\cos\theta)^{n-p-2k} = r^{2k}\,(r\cos\theta)^{n-p-2k}\,r^p\sin^p\theta\sin p\phi.$$

Here the first two factors are polynomials in x, y, z; and for the last factor, if p is odd say and $= 2s+1$,

$$\sin p\phi = \sin\phi\,\{\text{sum of terms of the form } (\sin\phi)^{2s-2m}\},$$

and

$$r^p \sin^p\theta\sin p\phi = r\sin\theta\sin\phi\,\{\text{sum of terms of the form}$$
$$(r\sin\theta)^{2m}\,(r\sin\theta\sin\phi)^{2s-2m}\}$$
$$= y\,\{\text{sum of terms of the form } (x^2+y^2)^m\,y^{2s-2m}\}.$$

Similarly if p is even; and thus $r^n X_n{}^p \sin p\phi$ is expressed as a polynomial in x, y, z, and H_n is expressed in terms of $2n+1$ linearly independent polynomials.

6·44. The integral $\int S_n P_n \, d\omega$. *To prove that, if S_n be any surface harmonic of order n of the angular co-ordinates θ, ϕ and P_n the Legendre function of the same order having (θ',ϕ') as its axis, then* $\int S_n P_n \, d\omega = \dfrac{4\pi}{2n+1}\, S_n{}'$, *where the integration is over the whole surface of the unit sphere and $S_n{}'$ is the same function of θ', ϕ' that S_n is of θ, ϕ.*

Take the axis of P_n as the axis from which θ is measured so that $\mu = \cos\theta$. Then

$$d\omega = \sin\theta\,d\theta\,d\phi = -d\mu\,d\phi,$$

and

$$\int S_n P_n\,d\omega = \int_{-1}^{1}\int_{0}^{2\pi}\left\{A_0\,X_n{}^0 + \sum_{1}^{n}(A_p X_n{}^p\cos p\phi\right.$$

$$\left. + B_p X_n{}^p\sin p\phi)\right\}P_n\,d\mu\,d\phi,$$

where A_p, B_p are constants.

But $\int_0^{2\pi}\cos p\phi\,d\phi = 0$, $\int_0^{2\pi}\sin p\phi\,d\phi = 0$ and $X_n{}^0 = P_n$, so that

$$\int S_n P_n\,d\omega = 2\pi A_0\int_{-1}^{1}P_n{}^2\,d\mu = \frac{4\pi A_0}{2n+1},$$

where A_0 is constant.

But from **6·42** (4), when $\mu = 1$, $X_n{}^p$ vanishes save when $p = 0$. Hence, when $\mu = 1$,

$$S_n = A_0 X_n{}^0 = A_0 P_n.$$

But, when $\mu = 1$, $P_n = 1$, so that the constant A_0 is the value of S_n at the point on the sphere for which θ is zero, i.e. on the axis (θ', ϕ'). Hence $A_0 = S_n{}'$ and

$$\int S_n P_n\,d\omega = \frac{4\pi}{2n+1}\,S_n{}' \quad\ldots\ldots\ldots\ldots(1).$$

6·45. Expansion of a function in surface harmonics. A polynomial in $\cos\theta$, $\sin\theta\cos\phi$, $\sin\theta\sin\phi$ can be expressed in terms of surface harmonics. For this purpose we use the following theorem: *If H_n is a solid harmonic of degree n, then*

$$\nabla^2(r^m H_n) = m(2n+m+1)r^{m-2}H_n,$$

where $r^2 = x^2 + y^2 + z^2$.

We have $\quad \dfrac{\partial}{\partial x}(r^m H_n) = mr^{m-2}x H_n + r^m\dfrac{\partial H_n}{\partial x}$

and

$$\frac{\partial^2}{\partial x^2}(r^m H_n) = m(m-2)r^{m-4}x^2 H_n + mr^{m-2}H_n$$

$$+ 2mr^{m-2}x\frac{\partial H_n}{\partial x} + r^m\frac{\partial^2 H_n}{\partial x^2}.$$

By adding the similar equations in y and z, we get

$$\nabla^2 (r^m H_n) = m(m+1) r^{m-2} H_n + 2m r^{m-2} \left(x \frac{\partial H_n}{\partial x} + y \frac{\partial H_n}{\partial y} + z \frac{\partial H_n}{\partial z} \right).$$

But since H_n is homogeneous and of degree n

$$x \frac{\partial H_n}{\partial x} + y \frac{\partial H_n}{\partial y} + z \frac{\partial H_n}{\partial z} = n H_n,$$

therefore $\qquad \nabla^2 (r^m H_n) = m(2n+m+1) r^{m-2} H_n$ ....(1).

Now let $f_n(x,y,z)$ be a homogeneous polynomial of degree n in x, y, z, and suppose that

$$f_n (x,y,z) = H_n + r^2 H_{n-2} + r^4 H_{n-4} + \ldots \quad \ldots\ldots(2),$$

where H_n, H_{n-2}, ... are solid harmonics of the degrees indicated by the suffixes and the series ends with $r^{n-1} H_1$ or $r^n H_0$ according as n is odd or even. By using the result (1) we can shew how to determine the H's.

Thus by applying (1) successively, we get

$$\left.\begin{aligned}
\nabla^2 f_n &= 2(2n-1) H_{n-2} + 4(2n-3) r^2 H_{n-4} \\
&\qquad + 6(2n-5) r^4 H_{n-6} + \ldots \\
\nabla^4 f_n &= 2 \cdot 4 (2n-3)(2n-5) H_{n-4} \\
&\qquad + 4 \cdot 6 (2n-5)(2n-7) r^2 H_{n-6} + \ldots \\
&\ldots\ldots\ldots\ldots\ldots\ldots\ldots\ldots\ldots\ldots\ldots\ldots\ldots\ldots\ldots\ldots\ldots\ldots\ldots \\
&\ldots\ldots\ldots\ldots\ldots\ldots\ldots\ldots\ldots\ldots\ldots\ldots\ldots\ldots\ldots\ldots\ldots\ldots\ldots \\
\nabla^{n-1} f_n &= (n-1)(n+2)(n-3)n \ldots 2 \cdot 5 H_1 \quad (n \text{ odd}) \\
\text{or } \nabla^n f_n &= n(n+1)(n-2)(n-1) \ldots 2 \cdot 3 H_0 \quad (n \text{ even})
\end{aligned}\right\} \;\ldots(3).$$

From the last of equations (3) we can find H_1 or H_0 according as n is odd or even, then from the preceding equation we can find H_3 or H_2 and so on until the H's up to H_{n-2} are determined and then H_n is given by (2).

By changing into polar co-ordinates, writing $r \cos \theta$, $r \sin \theta \cos \phi$, $r \sin \theta \sin \phi$ for x, y, z, and then dividing by r^n, it follows that any polynomial in $\cos \theta$, $\sin \theta \cos \phi$, $\sin \theta \sin \phi$ can in this way be expressed in a series of surface harmonics.

6·5. Surface density in terms of surface harmonics. If as in **6·2** we find the potential at P due to a number of particles situated on the sphere of radius a, we get results of the form

$$V_1 = \sum_0^\infty \frac{r^n}{a^{n+1}} U_n, \quad r < a \quad \dots\dots\dots\dots(1)$$

and

$$V_2 = \sum_0^\infty \frac{a^n}{r^{n+1}} U_n, \quad r > a \quad \dots\dots\dots\dots(2),$$

where U_n denotes the sum of a finite number of surface harmonics (one for each particle) and therefore itself a surface harmonic.

We infer that the same result would be true for any arbitrary distribution of surface density on the sphere, since every particle of the matter gives rise to potential functions of the form stated.

Conversely, we can assume (1) and (2) to give the potential of a certain distribution of matter and proceed to find it. By hypothesis U_n is harmonic, so that $\nabla^2 V_1 = 0$ and $\nabla^2 V_2 = 0$; the matter therefore resides on the surface of the sphere, and its surface density σ is given by

$$4\pi\sigma = \left(\frac{\partial V_1}{\partial r} - \frac{\partial V_2}{\partial r} \right)_{r=a}$$

or

$$\sigma = \sum_0^\infty \frac{2n+1}{4\pi a^2} U_n \quad \dots\dots\dots\dots(3).$$

It follows that if we accept the physical argument that an arbitrary distribution of surface density on the surface of a sphere produces the same kind of field of potential as an aggregate of particles distributed over the sphere, viz. the potential given by (1) and (2), then the arbitrary surface density is expressible in the form (3). This implies that an arbitrary function of the two variables θ, ϕ on the surface of a sphere is expressible in a series of surface harmonics. The validity of this development in series has been discussed by many writers; a bibliography is given in *Encyclopédie des Sciences Mathématiques.**

* Tome II, vol. **v**, p. 176. *V.* also H. Poincaré, *Figures d'équilibre*, p. 52 [1902].

6·51. Reverting to the case in which the field is due to a single unit particle at a point M on the sphere $r=a$ (6·2), it is of interest to verify that the formula 6·5 (3) for the surface density gives zero surface density except at the point M.

Taking the values of V from 6·2 (3) and (4), the formula 6·5 (3) is easily seen to give

$$4\pi a^2\sigma=\sum_0^\infty (2n+1)P_n,$$

and it is convenient to take for P_n the formula 6·21 (2) putting $r=a$, so that

$$4\pi a^2\sigma=\left[\sum_0^\infty \frac{(-1)^n(2n+1)a}{n!}\left(f\frac{\partial}{\partial x}+g\frac{\partial}{\partial y}+h\frac{\partial}{\partial z}\right)^n\frac{1}{r}\right]_{r=a},$$

where f, g, h are the co-ordinates of M.

This is equivalent to

$$4\pi a\sigma=\sum_0^\infty\left[\frac{-2n(-1)^{n-1}}{n!}\left(f\frac{\partial}{\partial x}+g\frac{\partial}{\partial y}+h\frac{\partial}{\partial z}\right)\left(f\frac{\partial}{\partial x}+g\frac{\partial}{\partial y}+h\frac{\partial}{\partial z}\right)^{n-1}\frac{1}{r}\right.$$
$$\left.+\frac{(-1)^n}{n!}\left(f\frac{\partial}{\partial x}+g\frac{\partial}{\partial y}+h\frac{\partial}{\partial z}\right)^n\frac{1}{r}\right]_{r=a}.$$

But $\quad\sum_0^\infty\frac{n(-1)^{n-1}}{n!}\left(f\frac{\partial}{\partial x}+...\right)^{n-1}\frac{1}{r}=\sum_0^\infty\frac{(-1)^n}{n!}\left(f\frac{\partial}{\partial x}+...\right)^n\frac{1}{r}$

$$=\frac{1}{\{(x-f)^2+(y-g)^2+(z-h)^2\}^{\frac12}}.$$

Therefore

$$4\pi a\sigma=\left[\left\{-2\left(f\frac{\partial}{\partial x}+g\frac{\partial}{\partial y}+h\frac{\partial}{\partial z}\right)+1\right\}\frac{1}{\{(x-f)^2+...\}^{\frac12}}\right]_{r=a}$$
$$=\left[\frac{2f(x-f)+...+(x-f)^2+...}{\{(x-f)^2+...\}^{\frac32}}\right]_{r=a}$$
$$=\left[\frac{x^2+y^2+z^2-f^2-g^2-h^2}{\{(x-f)^2+...\}^{\frac32}}\right]_{r=a}$$
$$=\frac{OP^2-OM^2}{PM^3},$$

and this is zero except at M.

6·6. Potential of a thin spherical shell.

In accordance with 6·5 we assume that any distribution of surface density on the surface of a sphere may be represented by a series of surface harmonics. We may therefore represent the surface density at any point $M(a,\theta',\phi')$ by

$$\sigma=S_0+S_1+S_2+...+S_n+... \quad............(1).$$

If P is the point (r, θ, ϕ), the potential at P due to an element σdS of the shell is (as in **6·2** and **6·29**) given by

$$
\left.
\begin{aligned}
V_1 &= \frac{\sigma dS}{a}\left\{1 + \frac{r}{a}P_1 + \frac{r^2}{a^2}P_2 + \ldots + \frac{r^n}{a^n}P_n + \ldots\right\}, \quad r < a \\
\text{and} \quad V_2 &= \frac{\sigma dS}{r}\left\{1 + \frac{a}{r}P_1 + \frac{a^2}{r^2}P_2 + \ldots + \frac{a^n}{r^n}P_n + \ldots\right\}, \quad r > a
\end{aligned}
\right\}
$$
$$\ldots\ldots(2).$$

And the potential at P due to the whole sphere is got by substituting the series (1) for σ and integrating (2) over the sphere, and by writing $a^2 d\omega$ for dS and using the theorem **6·44**, we find that

$$
\left.
\begin{aligned}
V_1 &= 4\pi a\left\{S_0' + \frac{1}{3}\frac{r}{a}S_1' + \frac{1}{5}\frac{r^2}{a^2}S_2' + \ldots + \frac{1}{2n+1}\frac{r^n}{a^n}S_n' + \ldots\right\}, \; r < a \\
\text{and} & \\
V_2 &= \frac{4\pi a^2}{r}\left\{S_0' + \frac{1}{3}\frac{a}{r}S_1' + \frac{1}{5}\frac{a^2}{r^2}S_2' + \ldots + \frac{1}{2n+1}\frac{a^n}{r^n}S_n' + \ldots\right\}, \; r > a
\end{aligned}
\right\}
$$
$$\ldots\ldots(3),$$

where S_0', S_1', S_2', etc. denote the values of the surface harmonics S_0, S_1, S_2, etc. at the point on the sphere where the radius through P intersects it.

It is easy to verify that the formula of **3·7**

$$4\pi\sigma = \left\{\frac{\partial V_1}{\partial r} - \frac{\partial V_2}{\partial r}\right\}_{r=a} \qquad \ldots\ldots\ldots\ldots\ldots(4)$$

leads to the expression (1) for the surface density.

6·61. Examples. (i) *The density of a thin spherical shell of radius a is given by $\sigma = \lambda z^2$, where z denotes distance from a diametral plane. Find the potential.*

We must first express the density in terms of harmonics. We have

$$
\begin{aligned}
\sigma &= \lambda z^2 = \lambda a^2 \cos^2\theta \\
&= \tfrac{1}{3}\lambda a^2(P_0 + 2P_2) \qquad\qquad\ldots\ldots\ldots\ldots\ldots\ldots(1),
\end{aligned}
$$

since $P_0 = 1$ and $P_2 = \tfrac{1}{2}(3\cos^2\theta - 1)$. We can now either write down the potential from **6·6** (3), or we may assume a series of harmonics for V and find the coefficients by using **6·6** (4). Thus, let

$$
\left.
\begin{aligned}
V_1 &= c_0 P_0 + c_2\frac{r^2}{a^2}P_2, \quad r < a \\
\text{and} \quad V_2 &= c_0\frac{a}{r}P_0 + c_2\frac{a^3}{r^3}P_2, \quad r > a
\end{aligned}
\right\}
\qquad \ldots\ldots\ldots\ldots\ldots(2).
$$

Then substitute for V_1, V_2 and σ in the relation

$$\left(\frac{\partial V_1}{\partial r}-\frac{\partial V_2}{\partial r}\right)_{r=a}=4\pi\sigma,$$

and we get $\dfrac{c_0}{a}P_0+\dfrac{5c_2}{a}P_2=\frac{4}{3}\pi\lambda a^2\,(P_0+2P_2)$ ....(3).

Since this relation must hold good at all points on the sphere, i.e. for all values of θ, the coefficients of P_0 and P_2 must be the same on both sides, so that

$$c_0=\tfrac{4}{3}\pi\lambda a^3 \quad\text{and}\quad c_2=\tfrac{2}{5}\cdot\tfrac{4}{3}\pi\lambda a^3$$

and
$$\left.\begin{aligned}
V_1&=\tfrac{4}{3}\pi\lambda a^3\left(1+\frac{2}{5}\frac{r^2}{a^2}P_2\right), \quad r<a\\[4pt]
V_2&=\tfrac{4}{3}\pi\lambda a^3\left(\frac{a}{r}+\frac{2}{5}\frac{a^3}{r^3}P_2\right), \quad r>a
\end{aligned}\right\} \quad\text{...............(4).}$$

We remark that in adopting this latter method we only insert in the series (2) terms to correspond to the harmonics that occur in (1). If any additional terms were inserted, the relation (3) would serve to shew that their coefficients were all zero.

Further, the total mass M of the shell is given by

$$M=\int_0^\pi \lambda a^2\cos^2\theta\,.\,2\pi a^2\sin\theta\,d\theta=\tfrac{4}{3}\pi\lambda a^4,$$

and since $2r^2P_2=r^2\,(3\cos^2\theta-1)=2z^2-x^2-y^2,$

therefore the expressions for the potential may be written

$$\left.\begin{aligned}
V_1&=\frac{M}{a}+\frac{M}{5a^3}(2z^2-x^2-y^2), \quad r<a\\[4pt]
V_2&=\frac{M}{r}+\frac{M}{5}\frac{a^2}{r^5}(2z^2-x^2-y^2), \quad r>a
\end{aligned}\right\} \quad\text{............(5).}$$

(ii) *The density of a solid sphere of radius a is given by $\rho=\lambda xyz$. Find the potential at an external point.*

Consider a thin shell of radius a' and thickness da'. We may regard it as a layer of surface density

$$\sigma=\rho\,da'=\lambda xyz\,da'   \quad\text{........................(1).}$$

Now xyz is a harmonic function of the third degree, so that by analogy from 6·6 this and its associated harmonic xyz/r^7 [6·13] are the only harmonics which can occur in the expressions for the potential. We therefore assume that for the potentials of the shell

$$V_1=Cxyz, \quad r<a'$$

and $V_2=Ca'^7\dfrac{xyz}{r^7}, \quad r>a',$

arranging the constants so that $V_1=V_2$ when $r=a'$, V_1 is finite at the origin and $V_2\to0$ as $r\to\infty$.

Then substituting in

$$\left(\frac{\partial V_1}{\partial r} - \frac{\partial V_2}{\partial r}\right)_{r=a'} = 4\pi\sigma,$$

we get
$$C\left\{\frac{\partial xyz}{\partial r} - \frac{\partial xyz}{\partial r} + 7\frac{xyz}{a'}\right\} = 4\pi\lambda xyz\, da',$$

so that
$$C = \tfrac{4}{7}\pi\lambda a'\, da'.$$

Hence the external potential of this shell is $\frac{4}{7}\pi\lambda a'^8 da' . \dfrac{xyz}{r^7}$. And by integrating with respect to a' between limits 0 and a, we get for the potential of the sphere $\frac{4}{63}\pi\lambda a^9 \dfrac{xyz}{r^7}$.

(iii) *A mass M is distributed over a spherical surface of radius a so that the surface density at any point is proportional to $y^2z^2 + z^2x^2 + x^2y^2$, where the origin is at the centre of the sphere. Find the potential inside and outside the sphere.*

It is necessary in the first place to express the surface density in a series of harmonic functions. Proceeding as in 6·45, we put

$$y^2z^2 + z^2x^2 + x^2y^2 = f(x,y,z) = H_4 + r^2 H_2 + r^4 H_0 \quad \ldots\ldots(1),$$

where H_4, H_2, H_0 are harmonics of degree 4, 2, 0.

Then
$$\nabla^2 f = 2.7 H_2 + 4.5 r^2 H_0 \quad \ldots\ldots\ldots\ldots\ldots(2)$$

and
$$\nabla^4 f = 4.5.2.3 H_0 \quad \ldots\ldots\ldots\ldots\ldots(3).$$

But, by differentiation,

$$\nabla^2 f = 4(x^2 + y^2 + z^2)$$

and
$$\nabla^4 f = 24.$$

Therefore, from (3), (2) and (1),

$$H_0 = \tfrac{1}{5}, \quad H_2 = 0$$

and
$$H_4 = y^2z^2 + z^2x^2 + x^2y^2 - \tfrac{1}{5}r^4.$$

We may therefore express the surface density in the form

$$\sigma = \lambda(y^2z^2 + z^2x^2 + x^2y^2 - \tfrac{1}{5}r^4) + \tfrac{1}{5}\lambda a^4,$$

and assume expressions for the potential containing the same harmonics, viz.

$$V_1 = A + B(y^2z^2 + z^2x^2 + x^2y^2 - \tfrac{1}{5}r^4), \qquad r < a$$

and
$$V_2 = A\frac{a}{r} + B\frac{a^9}{r^9}(y^2z^2 + z^2x^2 + x^2y^2 - \tfrac{1}{5}r^4), \quad r > a.$$

Then, by substituting in the relation

$$\left(\frac{\partial V_1}{\partial r} - \frac{\partial V_2}{\partial r}\right)_{r=a} = 4\pi\sigma,$$

we get $\quad \dfrac{A}{a} + \dfrac{9B}{a}(y^2z^2 + z^2x^2 + x^2y^2 - \frac{1}{5}a^4)$

$$= 4\pi\lambda\,(y^2z^2 + z^2x^2 + x^2y^2 - \tfrac{1}{5}a^4) + \tfrac{4}{5}\pi\lambda a^4,$$

so that $\qquad\qquad A = \tfrac{4}{5}\pi\lambda a^5 \quad$ and $\quad B = \tfrac{4}{9}\pi\lambda a.$

But $M = \displaystyle\int \sigma \, dS$, integrated over the sphere,

$\qquad = \tfrac{1}{5}\displaystyle\int \lambda a^4 \, dS$, since the integral of a harmonic is zero,

$\qquad = \tfrac{4}{5}\pi\lambda a^6.$

Hence it follows that

$$V_1 = M\left\{\frac{1}{a} + \frac{5}{9a^5}(y^2z^2 + z^2x^2 + x^2y^2 - \tfrac{1}{5}r^4)\right\}$$

and $\qquad V_2 = M\left\{\dfrac{1}{r} + \dfrac{5}{9}\dfrac{a^4}{r^9}(y^2z^2 + z^2x^2 + x^2y^2 - \tfrac{1}{5}r^4)\right\}.$

6·7. Nearly spherical bodies. Suppose that the strata of equal density in a body are nearly spherical, so that the equation of a surface of equal density is

$$r = a\,(1 + \Sigma C_n S_n),$$

where the coefficients C_n are so small that their squares and products can be neglected.

The volume contained by such a surface is $\tfrac{1}{3}\displaystyle\int r^3 \, d\omega$ [**1·42**], which

$$= \tfrac{1}{3}\int (a^3 + 3a^3 \Sigma C_n S_n)\, d\omega$$

$$= \tfrac{4}{3}\pi a^3, \quad \text{since } \int S_n \, d\omega = 0, \ [\mathbf{6\cdot41}],$$

$$= \text{vol. of a sphere of radius } a.$$

We note as special cases:

　(i) The surface $\quad r = a + \epsilon P_1,$

or $\qquad\qquad\qquad r = a + \epsilon\cos\theta.$

When ϵ is small enough, this is a sphere of radius a with its centre C at a small distance ϵ from the origin.

　(ii) The surface $r = a + \epsilon P_2,$

or $\qquad\qquad\qquad r^2 = a^2 + 2a\epsilon P_2.$

To the first order in powers of ϵ, this is the same as

$$r^2 = a^2 + 2\frac{r^2}{a}\epsilon P_2,$$

or

$$r^2 = a^2 + \frac{r^2}{a}\epsilon(3\cos^2\theta - 1).$$

In Cartesian co-ordinates this can be reduced to

$$\frac{x^2+y^2}{(a-\frac{1}{2}\epsilon)^2} + \frac{z^2}{(a+\epsilon)^2} = 1,$$

so that the surface is a spheroid of semi-axes $a - \frac{1}{2}\epsilon$, $a+\epsilon$.

We also notice that the normal to such a surface makes with the radius vector r an angle χ which is at most of order ϵ. So that for differentiation along the normal we have

$$\frac{\partial V}{\partial n} = \lambda\frac{\partial V}{\partial r} + \mu\frac{\partial V}{r\,\partial\theta} + \nu\frac{\partial V}{r\sin\theta\,\partial\phi},$$

where $\lambda = \cos\chi = 1$ to the first order, while μ, ν are of higher order than the first in ϵ. Consequently, to the first order in ϵ, we have

$$\partial V/\partial n = \partial V/\partial r.$$

6·71. Potential of a nearly spherical body.

Consider a body stratified as in **6·7**. The surface

$$r = a(1 + \Sigma C_n S_n) \qquad \dots\dots\dots\dots(1)$$

may be regarded as the boundary of a solid sphere of radius a and unit density surrounded by a layer of surface density $a\Sigma C_n S_n$.

The potentials inside and outside the sphere are, from **6·6**,

$$\left.\begin{aligned}
U_1 &= \tfrac{2}{3}\pi(3a^2 - r^2) + 4\pi\Sigma\frac{C_n}{2n+1}\frac{r^n}{a^{n-2}}S_n', \quad r<a\\
\text{and}\quad U_2 &= \frac{4}{3}\frac{\pi a^3}{r} + 4\pi\Sigma\frac{C_n}{2n+1}\frac{a^{n+3}}{r^{n+1}}S_n', \qquad r>a
\end{aligned}\right\} \dots(2).$$

If we regard a as variable, the differentials of U_1, U_2 will give the potentials inside and outside the shell whose parameters are a and $a+da$. The density ρ of the corresponding stratum of the given body is by hypothesis a function of a, and if c, c' are the values of a on the inner and outer boundaries of the body, the potentials inside and outside the shell are given by

$$V_1 = 4\pi \int_c^{c'} \rho \left\{ a + \frac{d}{da} \Sigma \frac{C_n S_n'}{2n+1} \frac{r^n}{a^{n-2}} \right\} da$$

and

$$V_2 = \frac{4\pi}{r} \int_c^{c'} \rho \left\{ a^2 + \frac{d}{da} \Sigma \frac{C_n S_n'}{2n+1} \frac{a^{n+3}}{r^n} \right\} da$$

$$\left. \right\} \quad \ldots \ldots (3).$$

The potential in the substance of the body, on the surface of parameter b say, will be found by adding together the first integral between limits b and c' and the second taken from c to b.

6·72. Example. *The form of a nearly spherical uniform gravitating solid of mass M is given by*

$$r = a(1 + \epsilon lmn),$$

where l, m, n are the direction cosines of r. Prove that neglecting ϵ^2 the exterior potential is

$$\gamma M \left\{ \frac{1}{r} + \frac{3}{7} \epsilon \frac{a^3 lmn}{r^4} \right\}.$$

If the solid is surrounded by a thin layer of water just sufficient everywhere to cover its surface, shew that, neglecting the self-gravitation of the water, the maximum depth of the water is $\dfrac{8\epsilon a}{21\sqrt{3}}$.   [C. 1931]

Since ϵ is small we may regard the solid as formed of a sphere of radius a and uniform density ρ surrounded by a thin layer of surface density $\sigma = a\epsilon\rho lmn$; and by [6·7] $\frac{4}{3}\pi a^3 \rho = M$, the whole mass.

Also, since xyz or $r^3 lmn$ is a solution of Laplace's equation, therefore [6·13] xyz/r^7 or lmn/r^4 is also a solution.

The potentials inside and outside the sphere due to the surface layer may be taken to be of the form

$$V_1 = A \frac{r^3}{a^3} lmn, \quad \text{and} \quad V_2 = A \frac{a^4}{r^4} lmn,$$

so that V_1 is finite at the centre, V_2 vanishes at infinity and $V_1 = V_2$ at the surface. Then from 3·7 and 6·7,

$$\left(\frac{\partial V_1}{\partial r} - \frac{\partial V_2}{\partial r} \right)_{r=a} = 4\pi\gamma\sigma = 4\pi\gamma a\epsilon\rho lmn.$$

Therefore $7A = 4\pi\gamma a^2 \epsilon \rho = 3\gamma M\epsilon/a,$

and the external potential due to the whole body is

$$V = \gamma M \left(\frac{1}{r} + \frac{3}{7} \epsilon \frac{a^3}{r^4} lmn \right) \quad \ldots\ldots\ldots\ldots\ldots\ldots(1).$$

The free surface of the water must be an equipotential surface because in equilibrium the surfaces of equal pressure are at right angles to the resultant force.

Hence the free surface is given by an equation

$$\frac{1}{r} + \frac{3}{7}\,\epsilon\,\frac{a^3}{r^4}\,lmn = C \qquad \dots\dots\dots\dots\dots\dots(2),$$

where C is a constant to be determined from the fact that the water only just covers the solid. The surface of the water will touch the solid on its protuberances, i.e. where the surface $r = a\,(1 + \epsilon lmn)$ has its maximum radii; and the water will be deepest over the hollows on the surface of the solid, i.e. where $r = a\,(1 + \epsilon lmn)$ has its minima.

It is easy to shew that lmn has maxima and minima values subject to $l^2 + m^2 + n^2 = 1$, where $l = \pm\dfrac{1}{\sqrt{3}}$, $m = \pm\dfrac{1}{\sqrt{3}}$, $n = \pm\dfrac{1}{\sqrt{3}}$, maxima when the signs are chosen to make lmn positive, and minima when negative.

Hence C is to be found by substituting $\dfrac{1}{3\sqrt{3}}$ for lmn and $a\left(1 + \dfrac{\epsilon}{3\sqrt{3}}\right)$ for r; so that

$$C = \frac{1}{a}\left(1 - \frac{\epsilon}{3\sqrt{3}}\right) + \frac{\epsilon}{7\sqrt{3}\,a} = \frac{1}{a}\left(1 - \frac{4\epsilon}{21\sqrt{3}}\right),$$

and the free surface is given by

$$\frac{1}{r} + \frac{3}{7}\,\epsilon\,\frac{a^3}{r^4}\,lmn = \frac{1}{a}\left(1 - \frac{4\epsilon}{21\sqrt{3}}\right) \qquad \dots\dots\dots\dots\dots(3).$$

A minimum radius of the body is obtained by putting $lmn = -\dfrac{1}{3\sqrt{3}}$, so that the minimum radius is $r = a\left(1 - \dfrac{\epsilon}{3\sqrt{3}}\right)$, and the radius of the free surface in this position is given by

$$\frac{1}{r} - \frac{1}{7\sqrt{3}}\,\frac{\epsilon a^3}{r^4} = \frac{1}{a}\left(1 - \frac{4\epsilon}{21\sqrt{3}}\right).$$

To the first power of ϵ, this gives

$$\frac{1}{r} = \frac{1}{a}\left(1 - \frac{4\epsilon}{21\sqrt{3}} + \frac{\epsilon}{7\sqrt{3}}\right) = \frac{1}{a}\left(1 - \frac{\epsilon}{21\sqrt{3}}\right),$$

or

$$r = a\left(1 + \frac{\epsilon}{21\sqrt{3}}\right).$$

So the greatest depth of the water

$$= a\left(1 + \frac{\epsilon}{21\sqrt{3}}\right) - a\left(1 - \frac{\epsilon}{3\sqrt{3}}\right)$$

$$= 8\epsilon a/21\sqrt{3}.$$

6·8. Clairaut's Theorem.* Variation of gravity. As the result of the measurement of meridian arcs there is reason to suppose that the earth is a slightly oblate spheroid. If a, c are the major and minor semi-axes of such a spheroid and $c = a(1 - \epsilon)$, ϵ is called the ellipticity of the spheroid. Neglecting the square of the ellipticity the equation of the surface is

$$\frac{x^2 + y^2}{a^2} + \frac{z^2}{a^2(1 - 2\epsilon)} = 1$$

or $$x^2 + y^2 + z^2(1 + 2\epsilon) = a^2$$

or $$r^2(1 + 2\epsilon\cos^2\theta) = a^2$$

or $$r = a(1 - \epsilon\cos^2\theta) \quad\ldots\ldots\ldots\ldots\ldots(1).$$

For the earth, ϵ is about $1/300$.

The earth rotates with a small angular velocity $\omega = 2\pi/24 . 60^2$. Bodies regarded as at rest relative to the earth are therefore subjected to a 'centrifugal force' as well as to the earth's attraction; so that if V denotes the potential of the earth's mass and X, Y, Z denote the whole force on a unit particle on the surface,

$$X = \frac{\partial V}{\partial x} + \omega^2 x, \quad Y = \frac{\partial V}{\partial y} + \omega^2 y, \quad Z = \frac{\partial V}{\partial z} \quad\ldots\ldots(2).$$

As by far the greater part of the earth's surface is fluid, it is reasonable to argue that the surface must be one of constant pressure, and from hydrostatical considerations this requires that $$X\,dx + Y\,dy + Z\,dz = 0 \ldots\ldots\ldots\ldots\ldots(3),$$

so that $$V + \tfrac{1}{2}\omega^2(x^2 + y^2) = C \quad\ldots\ldots\ldots\ldots\ldots(4)$$

over the surface, where C is a constant.

If E denotes the mass of the earth, its potential at a great distance is E/r, as a first approximation. We have to correct this because of the term $\epsilon\cos^2\theta$ in (1), and it will suffice for our purpose to add a solution of Laplace's equation containing $\cos^2\theta$, i.e. a term AP_2/r^3, where A is a small constant of order ϵ.

---

* Alexis Claude Clairaut (1713–1765). French mathematician.

Thus we take
$$V = \frac{E}{r} + \frac{AP_2}{r^3} \quad\text{......................(5)}$$

and determine A so that (4) is satisfied by (5) when r is given by (1); i.e.

$$\frac{E}{a}(1 + \epsilon \cos^2 \theta) + \frac{A}{a^3} \frac{3\cos^2\theta - 1}{2} + \tfrac{1}{2}\omega^2 a^2 (1 - \cos^2\theta) = C$$
$$\text{......(6).}$$

$r = a$ is a sufficient approximation in the second and third terms since A and ω^2 are of order ϵ.

Since (6) is to be satisfied for all values of θ, by equating to zero the coefficient of $\cos^2\theta$, we get

$$\frac{A}{a^3} = \frac{1}{3}\left(\omega^2 a^2 - \frac{2\epsilon E}{a}\right) \quad\text{.................(7),}$$

and the terms independent of $\cos^2\theta$ serve to determine C.

Hence
$$V = \frac{E}{r} + \frac{a^3}{r^3}\left(\tfrac{1}{2}\omega^2 a^2 - \frac{\epsilon E}{a}\right)(\cos^2\theta - \tfrac{1}{3}) \quad\text{......(8).}$$

The radius from the earth's centre to the point on its surface makes an angle θ with the axis Oz. The force of gravity g at the earth's surface is the resultant of the attraction and the centrifugal force and if it makes an angle ν with the radius vector the radial force is $-g\cos\nu$. But ν is so small that its square is negligible, so we have

$$g = -\frac{\partial}{\partial r}(V + \tfrac{1}{2}\omega^2 r^2 \sin^2\theta) \quad\text{at the surface,}$$

$$= \frac{E}{r^2} + \frac{3a^3}{r^4}\left(\tfrac{1}{2}\omega^2 a^2 - \frac{\epsilon E}{a}\right)(\cos^2\theta - \tfrac{1}{3}) - \omega^2 r \sin^2\theta,$$

where $r = a(1 - \epsilon \cos^2\theta)$ and ω^2 is of order ϵ.

This gives
$$g = \frac{E}{a^2}(1 + \epsilon) - \tfrac{3}{2}\omega^2 a + \left(\tfrac{5}{2}\omega^2 a - \frac{\epsilon E}{a^2}\right)\cos^2\theta \quad\text{......(9).}$$

If G denotes the value of g at the equator $\theta = \tfrac{1}{2}\pi$, and m denotes the ratio of the centrifugal force to gravity at the equator so that $\omega^2 a = mG$, we have

$$(1 + \tfrac{3}{2}m)G = (1 + \epsilon)\frac{E}{a^2} \quad\text{...............(10),}$$

so that $$E = (1 + \tfrac{3}{2}m - \epsilon)\, Ga^2 \quad\dots\dots\dots\dots(11)$$

and (9) becomes $$g = G\{1 + (\tfrac{5}{2}m - \epsilon)\cos^2\theta\}\quad\dots\dots\dots(12).$$

We observe that if G is found by means of a pendulum, (11) determines the mass of the earth.

The numerical value of m or $\omega^2 a/G$ is about $1/289$, so that it is of the same order as ϵ, and in the small term in (8) we may substitute $\omega^2 a = mG = mE/a^2$, so that the potential of the earth at an external point is given by

$$V = \frac{E}{r} + (\tfrac{1}{2}m - \epsilon)\frac{Ea^2}{r^3}(\cos^2\theta - \tfrac{1}{3}) \quad\dots\dots(13).$$

6·81. The Moon's attraction. The forces exerted by the moon on the earth are equal and opposite to the forces exerted by the earth on the moon, and are approximately the same as if the moon's mass M were collected at its centre of gravity. If r denotes the distance between the centres of the earth and moon, and θ the moon's north-polar distance, the earth's attraction at the centre of the moon is composed of a radial attraction $P = -\dfrac{\partial V}{\partial r}$ and a force $Q = \dfrac{1}{r}\dfrac{\partial V}{\partial \theta}$ tending to increase θ. Hence from **6·8** (13)

$$\left.\begin{aligned}
P &= \frac{E}{r^2} + 3\,(\tfrac{1}{2}m - \epsilon)\frac{Ea^2}{r^4}(\cos^2\theta - \tfrac{1}{3}) \\
Q &= -2\,(\tfrac{1}{2}m - \epsilon)\frac{Ea^2}{r^4}\sin\theta\cos\theta
\end{aligned}\right\} \quad\dots\dots(1).$$

The forces exerted on the moon are MP radially and MQ through the moon's centre. The force system exerted on the earth by the moon consists of equal and opposite forces MP, MQ acting through the earth's centre and a couple of moment MQr tending to cause rotation about an equatorial diameter, and since $\epsilon - \tfrac{1}{2}m$ is positive the sense of the couple is such as to decrease θ. The moment of this couple is

$$2\,(\epsilon - \tfrac{1}{2}m)\frac{Ea^2}{r^2}\sin\theta\cos\theta \quad\dots\dots\dots(2).$$

6·82. We can obtain some information by comparing the results of **6·8** and **6·81** with MacCullagh's expression for the potential in **4·6** and the couple deduced therefrom in **4·62**. We notice that the potential as given by **6·8** (13) is independent of the longitude, therefore when the attracting body is the earth MacCullagh's expression

$$V = \frac{E}{r} + \frac{A+B+C-3I}{2r^3} + \cdots$$

should be independent of the longitude. This is satisfied by taking a principal axis of the earth as the axis of rotation and writing $A = B$, so that $I = A \sin^2 \theta + C \cos^2 \theta$, and

$$V = \frac{E}{r} - \frac{3(C-A)}{2r^3} (\cos^2 \theta - \tfrac{1}{3}).$$

A comparison with **6·8** (13) now shews that

$$\frac{C-A}{Ea^2} = \tfrac{2}{3} (\epsilon - \tfrac{1}{2}m).$$

The same result follows from comparing the couples of **4·62** and **6·81**.

6·9. General solution of Laplace's equation. It has been shewn by Whittaker* that there is a general solution of Laplace's equation of the form

$$V = \int_{-\pi}^{\pi} f(z + ix \cos u + iy \sin u, u)\, du,$$

provided that f is such a function that differentiations with regard to x, y, z under the sign of integration are permissible; and it may be shewn that V can be expressed as a series of expressions of the types

$$\int_{-\pi}^{\pi} (z + ix \cos u + iy \sin u)^n \cos mu\, du,$$

$$\int_{-\pi}^{\pi} (z + ix \cos u + iy \sin u)^n \sin mu\, du,$$

* *Math. Ann.* LVII (1902), p. 333. See also Whittaker and Watson, *Modern Analysis*, ch. XVIII.

where n and m are integers such that $0 \leqslant m \leqslant n$; also that by changing to polar co-ordinates these may be reduced to solutions of the forms $X_n{}^p \cos p\phi$, $X_n{}^p \sin p\phi$ in the notation of 6·42, 6·43.

EXAMPLES

1. Find, in terms of μ, $P_4(\mu)$, $P_5(\mu)$, $P_6(\mu)$.

2. Prove that

$$\int_{-1}^{1} x^2 P_n{}^2(x)\, dx = \frac{\frac{1}{8}}{2n-1} + \frac{\frac{3}{4}}{2n+1} + \frac{\frac{1}{8}}{2n+3}.$$

[M. T. 1912]

3. Prove that $\displaystyle\int_{0}^{1} \{P_n(\mu)\}^2\, d\mu = \frac{1}{2n+1}$.

4. Prove that $P_m P_n$ can be expressed in the form

$$c_0 P_{m-n} + c_1 P_{m-n+1} + \ldots + c_{2n} P_{m+n},$$

where $m > n$ and the c's are constants.   [C. 1908]

5. Express $\dfrac{dP_n}{dx}$, $x\dfrac{dP_n}{dx}$ and $\dfrac{d^2 P_n}{dx^2}$ in the form

$$A + A_1 P_1 + A_2 P_2 + A_3 P_3 + \ldots$$

6. Prove that $\displaystyle\int_{-1}^{1} \left(\frac{dP_n}{d\mu}\right)^2 d\mu = n(n+1)$.

7. Prove that $\displaystyle\int_{-1}^{1} (1-\mu^2)\frac{dP_m}{d\mu}\frac{dP_n}{d\mu}\, d\mu = 0$, $m \neq n$,

and

$$\int_{-1}^{1} (1-\mu^2)\left(\frac{dP_n}{d\mu}\right)^2 d\mu = \frac{2n(n+1)}{2n+1}.$$

8. Shew that, if m and n are integers, the value of

$$\int_{-1}^{1} \mu P_n \frac{dP_m}{d\mu}\, d\mu$$

is either 0, 2 or $\dfrac{2n}{2n+1}$.   [M. T. 1908]

9. Prove that, if $f(\mu)$ is expressible in the form $\displaystyle\sum_{0}^{\infty} a_n P_n(\mu)$, then

$$\sum_{0}^{\infty} a_n x^n = \int_{-1}^{1} \frac{(1-x^2)f(\mu)\, d\mu}{(1-2\mu x + x^2)^{\frac{3}{2}}}.$$

10. Prove that $\displaystyle\int_{0}^{1} P_m(\mu) P_n(\mu)\, d\mu$ is equal to zero if m and n are unequal and both even or both odd integers, while if m is even and n is odd, its value is

$$\frac{(-1)^{\frac{1}{2}(m+n+1)}}{(m-n)(m+n+1)} \frac{1.3\ldots(m-1)}{2.4\ldots m} \frac{3.5\ldots n}{2.4\ldots(n-1)}. \qquad \text{[C. 1915]}$$

11. Explain why, if $F(\cos\theta)$ satisfies the equation

$$\frac{\partial^2 f}{\partial\theta^2} + \frac{1}{\sin^2\theta}\frac{\partial^2 f}{\partial\phi^2} + \cot\theta\frac{\partial f}{\partial\theta} + n(n+1)f = 0,$$

so also does $F(\lambda\cos\theta + \mu\sin\theta\cos\phi + \nu\sin\theta\sin\phi)$,

where $\lambda^2 + \mu^2 + \nu^2 = 1$.   [M. T. 1921]

12. Shew that

$$\int_0^1 x^2 P_{n+1}P_{n-1}\,dx = \frac{n(n+1)}{(2n-1)(2n+1)(2n+3)}.$$ [C. 1905]

13. If V_n is a homogeneous function of x, y, z, of degree n satisfying $\nabla^2 V = 0$, then

$$\left(A + Br^2 + \frac{C+Dr^2}{r^{2n+1}}\right)V_n \quad\text{satisfies } \nabla^2\nabla^2 V = 0.$$ [C. 1891]

14. Find the potentials of a surface distribution on a sphere in which the density at any point is proportional to the distance from a given plane.   [C. 1904]

15. A spherical surface of radius a is coated with matter of surface density proportional at each point to the cosine of the angle the radius vector from the centre to the point makes with the positive direction of the axis of x. Prove that the potential due to this matter at any external point is $\frac{4}{3}\pi\sigma a^3 x/r^3$, and at any internal point $\frac{4}{3}\pi\sigma x$; where the origin is at the centre and $\pm\sigma$ are the densities respectively at the points where the axis of x cuts the surface.   [C. 1890]

16. The surface density on a sphere of radius a is λxy. Find the potential inside and outside the sphere.

17. The surface of a sphere of radius a whose centre is at the origin is covered by a thin layer of attracting matter, the surface density at any point P being proportional to $\sin^2 POZ$. Find the potential at any point inside or outside the sphere.

Prove that inside the sphere the lines of force are given by $y/x = $ constant; $zx^2 = $ constant.   [London Univ. 1925]

18. Find the surface and volume distribution of matter for which

$$V = kr^p(3\cos^2\theta - 1) \quad\text{when } r \leqslant a,$$

and $V = ka^{p+3}r^{-3}(3\cos^2\theta - 1), \quad\text{when } r > a,$

k, p and a being constants.   [London Univ. 1936]

19. If the mass per unit area of a thin heterogeneous stratum of attracting matter placed on a sphere of radius a is $\rho z^3/a^3$, prove that the potential at a point within the sphere is

$$\tfrac{4}{35}\pi\rho(5z^3/a^2 - 3r^2 z/a^2 + 7z),$$

where $r^2 = x^2 + y^2 + z^2$, and find the potential at a point without the sphere.   [M.T. 1903]

20. Prove that Laplace's operator ∇^2 is invariant for any change of rectangular axes, and that a solid harmonic (or a surface harmonic) of degree n transforms into a solid harmonic (or a surface harmonic) of degree n in the new co-ordinates.

21. Prove that, if x be real and positive,

(1) $\displaystyle\int_{-1}^{1} \frac{P_n(\mu)\,d\mu}{(\cosh 2x - \mu)^{\frac{1}{2}}} = \frac{2^{\frac{3}{2}}}{(2n+1)\,e^{(2n+1)x}};$

(2) $\displaystyle\int_{-1}^{1} (\cosh 2x - \mu)^{\frac{1}{2}} P_n(\mu)\,d\mu$

$$= \frac{2^{\frac{1}{2}}}{(2n+1)\,(2n+3)\,e^{(2n+3)x}} - \frac{2^{\frac{1}{2}}}{(2n+1)\,(2n-1)\,e^{(2n-1)x}}.$$

[C. 1908]

22. If P_n is Legendre's coefficient of order n, prove that

$$x^2 \frac{d^2 P_n}{dx^2} = n\,(n-1)\,P_n + \sum_{r=1}^{n/2} (2n-4r+1)\,\{r\,(2n-2r+1)-2\}\,P_{n-2r}$$

when n is even, and that when n is odd the summation is from 1 to $\frac{1}{2}(n-1)$.

[M. T. 1904]

23. Express $xyz\,(x^3 + y^3 + z^3)$ in the form

$$V_6 + r^2 V_4 + r^4 V_2 + r^6 V_0,$$

where V_6, V_4, V_2, V_0 are solid harmonics of the orders indicated by the suffixes.

[London Univ. 1931]

24. Express in the form $V_6 + V_4 + V_2$, where V_n denotes a spherical solid harmonic of degree n, a function which satisfies Laplace's equation within a sphere $r = a$, is finite at the origin, and is equal to $x^3 y^2 z$ on the sphere $r = a$.

[London Univ. 1926]

25. Obtain a rational integral harmonic which has the value Az^4 at points on the surface of a unit sphere with its centre at the origin.

26. A thin layer of matter is placed on a sphere of radius a whose centre is O, the density at any point P being proportional to $\cos^4 AOP$, where OA is a fixed radius. Determine the potential at all points inside and outside the sphere.

[London Univ. 1927]

27. Matter is distributed in a thin layer over the surface of a sphere of radius a and centre at the origin, the density being proportional to $(x+y)^2$ and the whole mass being m. Shew that the potential at an external point is given by

$$\gamma \frac{m}{r}\left[1 + \frac{a^2}{10r^4}(x^2 + y^2 + 6xy - 2z^2)\right]$$

and find the value at an internal point.

[London Univ. 1938]

28. If the surface density at the point (x, y, z) on a sphere of radius a is $x^2 y - xy^2$, prove that the potential is

$$4\pi a (x - y) \left(\frac{xy}{7} - \frac{a^2}{15} + \frac{r^2}{35} \right)$$

at an internal point, and

$$4\pi a^6 (x - y) \left(\frac{a^2 xy}{7r^7} - \frac{1}{15r^3} + \frac{a^2}{35r^5} \right)$$

at an external point. [London Univ.]

29. The sphere $x^2 + y^2 + z^2 = a^2$ is covered by a thin layer of attracting matter, the total amount of matter being M and the surface density at any point (X, Y, Z) of the sphere being proportional to

$$Y^2 Z^2 + Z^2 X^2 + X^2 Y^2,$$

which may be written as

$$\tfrac{1}{6} (a^4 - X^4 - Y^4 - Z^4 + 3Y^2 Z^2 + 3Z^2 X^2 + 3X^2 Y^2).$$

Find the potential of the layer at any external point.

Shew that the attraction on a small mass m, placed outside the sphere on the line $x = y = z$ at a distance r from the origin, is

$$\frac{\gamma M m}{r^2} \left(1 + \frac{10a^4}{27r^4} \right). \qquad \text{[London Univ. 1925]}$$

30. If V_n is a rational integral harmonic function of degree n and $r^2 = x^2 + y^2 + z^2$, and U is a homogeneous function of x, y, z of degree $2n$,

shew (by expressing U in the form $\sum\limits_{p=0}^{n} r^{2n-2p} V_{2p}$, or otherwise) that

$$\iint U \, dS = \frac{4\pi a^{2n+2}}{(2n+1)!} (\nabla^2)^n \, U,$$

the integral being taken over the surface of the sphere $r = a$.

[M. T. 1939]

31. The density of a solid sphere of radius a, referred to axes through its centre, is $x^2 yz$. Shew that the potential of the sphere at an external point (x, y, z) is given by

$$V = \frac{4\pi a^2}{63} \left\{ \frac{H_4}{11} \cdot \left(\frac{a}{r} \right)^9 + \frac{H_2 a^2}{5} \cdot \left(\frac{a}{r} \right)^5 \right\},$$

where $H_2 \equiv yz, \quad H_4 \equiv 6x^2 yz - y^3 z - yz^3.$ [M. T. 1908]

32. The potential over a sphere of radius a is equal to $Axyz^2$, and is due to some surface distribution σ. Find the potential at an internal point, at an external point and the value of σ. [C. 1913]

33. Find the potential in all space when the potential at any point of the surface of a sphere is $y^2 z^2$. [C. 1906]

34. The density of a solid sphere of radius a at the point (x, y, z) is kz^3, where k is a constant. Prove that the potential at any point inside the sphere is

$$\frac{\pi\gamma kz}{35}\left[\tfrac{2}{9}\left(5z^2 - 3r^2\right)\left(9a^2 - 7r^2\right) + 7a^4 - 3r^4\right],$$

where $r^2 = x^2 + y^2 + z^2$. [London Univ. 1926]

35. The density of a solid sphere of radius a is $\rho_0 + \rho_1 S_2$, where ρ_0, ρ_1 are constants and S_2 is a spherical surface harmonic of the second degree with origin at the centre of the sphere. Prove that the potentials at internal and external points are

$$\tfrac{2}{3}\pi\gamma\rho_0\left(3a^2 - r^2\right) + \tfrac{4}{5}\pi\gamma\rho_1 r^2\left(\log\frac{a}{r}\right) S_2 + \tfrac{4}{25}\pi\gamma\rho_1 r^2 S_2$$

and

$$\tfrac{4}{3}\pi\gamma\rho_0\frac{a^3}{r} + \tfrac{4}{25}\pi\gamma\rho_1\frac{a^5}{r^3} S_2.$$

[London Univ. 1931]

36. The density of a solid sphere of radius a is proportional to $x^2 z^2$, the centre being the origin. Prove that the potential at an external point is

$$\gamma\frac{M}{r}\left\{1 + \frac{x^2 - 2y^2 + z^2}{9}\frac{a^2}{r^4} - \frac{7a^4}{99r^4} - \frac{5}{99}\left(x^2 - 2y^2 + z^2\right)\frac{a^4}{r^6} + \frac{35}{33}\frac{x^2 z^2 a^4}{r^8}\right\}.$$

[London Univ. 1926]

37. The matter of a heterogeneous solid sphere is arranged symmetrically about a diameter. It is found that the normal component of attraction due to the sphere at a point P on its surface where the radius makes angle θ with the diameter is $A + B\cos^4\theta$. Prove that the tangential component of attraction at P is

$$\tfrac{4}{35}B\sin\theta\cos\theta\left(2 + 7\cos^2\theta\right). \qquad \text{[C. 1907]}$$

38. A given mass is distributed in a thin layer over a spherical surface, symmetrically with respect to a diameter AB and to the corresponding diametral plane, so as to produce inside a field, such that the force at any point towards AB is proportional to the distance from it. Shew that the force parallel to AB is proportional to the distance from the corresponding diametral plane. Shew also that if the surface density in the diametral plane is zero, that at A is three times what it would be if the mass were uniformly distributed. [C. 1901]

39. Shew that the potential of a uniform circular ring of radius a and mass M is

$$\frac{M}{\pi}\int_0^\pi\left[a^2 + \{z^2 + i\left(x^2 + y^2\right)^{\frac{1}{2}}\cos\phi\}^2\right]^{-\frac{1}{2}}d\phi,$$

where the origin is at the centre and the axis of z along the axis of the ring.

40. Verify, by using the formulae of 6·23 for P_n, that the expressions V_1, V_2 of 6·32, for the potential of a circular disc, are equal when $r = a$, by proving them both equal to

$$2m \int_0^\pi [1 + \{\mu + \sqrt{(\mu^2 - 1)} \cos \phi\}^2]^{\frac{1}{2}} d\phi - 2\pi m a \mu.$$

41. Prove that, if the potential of a body symmetrical about an axis at a point on the axis at a distance z' from an origin is $\phi(z')$, then the potential at any point outside the body is

$$\frac{1}{\pi} \int_0^\pi \phi \{z + \sqrt{(z^2 + r^2)} \cos \psi\} d\psi,$$

where r is the distance of the point from the origin, and z the projection of r on the axis. [M. T. 1887]

42. Matter of mass M is distributed on the surface of a sphere whose centre is O and radius a, so that the density at any point is proportional to the square of its distance from an external point C, where $OC = b$. Prove that the potential at an external point P is

$$M \left\{ \frac{1}{r} - \frac{2a^2 b}{3(a^2 + b^2)} \frac{x}{r^3} \right\},$$

where $OP = r$ and $x = r \cos POC$. [C. 1897]

43. The density of a solid sphere at any point P varies inversely as OP, where O is an external point distant f from the centre. Shew that the potential at any external point is

$$\frac{3M}{r} \sum_0^\infty \left(\frac{a^2}{fr}\right)^n \frac{P_n(\cos \theta)}{(2n+1)(2n+3)},$$

where M is the mass of the sphere, a its radius, and r, θ polar coordinates whose origin is the centre and whose axis is the line from the centre to O. [M. T. 1909]

44. The density of a thin spherical shell, centre C, radius a and mass M, varies inversely as the distance from an external point O. Shew that the potential at any external point P is

$$\frac{1}{2} \frac{M}{a} \int_0^{a^2/f} \left\{ \frac{f}{x(r^2 - 2rx \cos \theta + x^2)} \right\}^{\frac{1}{2}} dx,$$

and at any internal point P is

$$\frac{1}{2} \frac{M}{a} \int_0^{r^2/f} \left\{ \frac{f}{x(r^2 - 2rx \cos \theta + x^2)} \right\}^{\frac{1}{2}} dx,$$

where $r = CP$, $f = CO$ and θ is the angle PCO. [M. T. 1895]

45. A disc of radius a is of a mass M and its surface density is proportional to the mth power of the distance from its centre ($m + 2 > 0$).

Shew that the potential at points whose distance (r) from the centre is greater than a is

$$\gamma M \left\{ \frac{1}{r} + (m+2) \sum_{n=1}^{\infty} (-1)^n \frac{1.3.5 \dots 2n-1}{2^n n! \, (m+2n+2)} \frac{a^{2n}}{r^{2n+1}} \, P_{2n} (\cos \theta) \right\}.$$

[London Univ.]

46. A thin circular disc of radius a is such that the mass per unit area at distance r from the centre is $k (a^2 - r^2)$. Find the attraction of the disc at any point on its axis.

Hence shew that the component attraction parallel to the axis of the disc at any point whose co-ordinates are r, θ, ϕ referred to the centre of the disc as origin and its axis as polar axis can be expressed in the form

$$2\pi k a^2 \sum_{1}^{\infty} (-1)^{n-1} \frac{1.3.5 \dots 2n-1}{(n+1)!} \left(\frac{a^2}{2r^2} \right)^n P_{2n-1} (\cos \theta),$$

if $r > a$. [M. T. 1920]

47. Prove that at a point in free space, where $r > a$, the potential of a homogeneous gravitating hemisphere of radius a and density ρ is

$$V = 2\pi\gamma\rho a^3 \left\{ \frac{1}{3r} + \frac{1}{8} \frac{a}{r^2} P_1 + \sum_{3}^{\infty} (-1)^n \frac{(2n-4)!}{2^{2n-2} n! \, (n-2)!} \frac{a^{2n-3} P_{2n-3}}{r^{2n-2}} \right\},$$

where r is the distance of the point from the centre of the base and the axis of harmonics is the axis of symmetry of the hemisphere drawn from the centre into the interior. [London Univ. 1931]

48. Shew that the potential of an oblate spheroid of density ρ and major and minor semi-axes a, c at large distances is approximately

$$4\pi\gamma\rho a^2 c \left[\frac{1}{1.3} \frac{P_0}{r} - \frac{a^2 - c^2}{3.5} \frac{P_2}{r^3} + \frac{(a^2 - c^2)^2}{5.7} \frac{P_4}{r^5} \right]. \qquad \text{[C. 1929]}$$

49. A uniform plane annulus of mass M is bounded by circles of radii a, b, large compared with the dimensions of a spheroid whose centre of gravity coincides with the centres of the circles. Shew that, if the normal to the plane of the annulus makes an angle θ with the axis of the spheroid, they exert on one another a couple of magnitude

$$\frac{3\gamma M (A - C)}{ab (a+b)} \sin \theta \cos \theta,$$

tending to increase θ, where A, A, C are the principal moments of inertia of the spheroid, and γ is the gravitation constant.

[M. T. 1931]

50. A uniform nearly spherical solid of density ρ has the surface $r = a (1 + \epsilon P_2)$ as its boundary. It is surrounded by liquid of volume $4\pi (b^3 - a^3)/3$ and uniform density σ. Shew that, provided the solid is completely covered with liquid, the equation of the free surface is $r = b (1 + \eta P_2)$, where

$$\eta = \frac{3 (\rho - \sigma) a^5 \epsilon}{b^2 \{5 (\rho - \sigma) a^3 + 2\sigma b^3\}}. \qquad \text{[M. T. 1936]}$$

Chapter VII

ATTRACTION OF ELLIPSOIDS

7·1. Homoeoids and Focaloids. A homoeoid is a shell bounded by two similar and similarly situated concentric ellipsoids. A **focaloid** is a shell bounded by two confocal ellipsoids.

Let a, b, c be the semi-axes of the internal surface of a *thin homoeoid* and $a + da$, $b + db$, $c + dc$ those of the external surface. Let O be the common centre and let a radius OPQ cut the surfaces in P, Q, where $OP = r$ and $OQ = r + dr$.

Since the figures are similar

$$\frac{dr}{r} = \frac{da}{a} = \frac{db}{b} = \frac{dc}{c} = k, \text{ say} \qquad \ldots\ldots\ldots\ldots(1).$$

Also the tangent planes at P, Q are parallel, and if p, $p + dp$ denote their distances from O and l, m, n the direction cosines of the perpendicular to them, we have

$$p^2 = a^2 l^2 + b^2 m^2 + c^2 n^2$$

and
$$\begin{aligned}(p + dp)^2 &= (a + da)^2 \, l^2 + (b + db)^2 \, m^2 + (c + dc)^2 \, n^2 \\ &= (1 + k)^2 \, (a^2 l^2 + b^2 m^2 + c^2 n^2), \quad \text{from (1)}, \\ &= (1 + k)^2 \, p^2.\end{aligned}$$

It follows that
$$dp/p = k \qquad \ldots\ldots\ldots\ldots\ldots\ldots(2);$$

and dp, the distance between two parallel tangent planes, is the thickness of the homoeoid at P.

The volume of the internal ellipsoid is $\frac{4}{3}\pi abc$ and that of the external ellipsoid is $\frac{4}{3}\pi \, (a + da) \, (b + db) \, (c + dc)$ or, from (1) $\frac{4}{3}\pi abc \, (1 + k)^3$. Therefore, neglecting higher powers of k, the volume of the thin homoeoid is $4\pi kabc$, when $k = da/a = \text{etc.}$

Thin homoeoids are said to be confocal when their inner boundaries are confocal.

For a **focaloid** for which the semi-axes of the internal and external ellipsoids are a, b, c and a', b', c' we have

$$a'^2 = a^2 + \lambda, \quad b'^2 = b^2 + \lambda, \quad c'^2 = c^2 + \lambda \ldots\ldots\ldots(3).$$

If the focaloid is thin and we take $a+da$, $b+db$, $c+dc$ as the semi-axes of the outer ellipsoid, we have from (3)

$$a\,da = b\,db = c\,dc = \tfrac{1}{2}\lambda \quad\dotsb\dotsb(4).$$

And if p, $p+dp$ are the distances from the centre of parallel tangent planes, and l, m, n direction cosines of the normal, we have

$$p^2 = a^2l^2 + b^2m^2 + c^2n^2$$

and

$$(p+dp)^2 = (a^2+\lambda)\,l^2 + (b^2+\lambda)\,m^2 + (c^2+\lambda)\,n^2,$$

so that

$$p\,dp = \tfrac{1}{2}\lambda \quad\dotsb\dotsb(5).$$

The volume of the thin focaloid is

$$\tfrac{4}{3}\pi\sqrt{\{(a^2+\lambda)\,(b^2+\lambda)\,(c^2+\lambda)\}} - \tfrac{4}{3}\pi abc$$

$$= \tfrac{2}{3}\pi\lambda abc\left(\frac{1}{a^2} + \frac{1}{b^2} + \frac{1}{c^2}\right) \quad\dotsb(6).$$

If ρ is the density of the matter and the thin shells are regarded as strata of matter condensed on the inner boundary, the mass per unit area is in either case $\rho\,dp$, and in the case of the homoeoid this is $k\rho p$, varying directly as p; but in the case of the focaloid it is $\tfrac{1}{2}\rho\lambda/p$, varying inversely as p.

7·2. Attraction of a homoeoid at an internal point.
With any point O inside the homoeoid as vertex, draw a cone of small solid angle $d\omega$ intersecting the homoeoid in frusta $PQQ'P'$, $RSS'R'$. If ρ is the density, the mass of an elementary slice of the cone of thickness dr at a distance r from O is $\rho r^2\,d\omega\,dr$, and its attraction at O is $\rho\,d\omega\,dr$. Consequently the attractions of the two intercepted frusta of the cone are $\rho\,d\omega\,PQ$ and $\rho\,d\omega\,RS$ in opposite 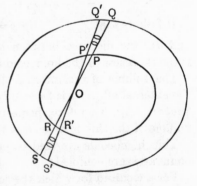 directions. But since the ellipsoids are similar, they make equal intercepts on any chord so that $PQ=RS$ and the

attractions of the frusta are equal and opposite. By taking cones in all directions round O, we see that the resultant attraction of the homoeoid at O is zero.

We appear to have neglected wedge-shaped elements at P and Q in integrating $\rho\, d\omega\, dr$ with regard to r, but the further integration with respect to $d\omega$ implies that $d\omega$ tends to zero when we regard the integral as the limit of a sum, so that these elements contribute nothing.

7·21. Potential of a homoeoid at an internal point. Since the attraction at an internal point is zero, therefore the potential is constant and equal to its value at the centre of the boundary surfaces. If we take a cone of small solid angle $d\omega$ with its vertex at the centre, the contribution to the potential of the frustum of the cone intercepted by the homoeoid is easily seen to be $\displaystyle\int_{r_1}^{r_2} \rho r\, d\omega\, dr = \tfrac{1}{2}\rho\,(r_2{}^2 - r_1{}^2)\, d\omega$, where r_1, r_2 are the radii of the inner and outer surfaces in the direction of the cone. If we take (a, b, c) and (ma, mb, mc) as the semi-axes of the outer and inner surfaces $r_1 = mr_2$ and the potential is

$$\tfrac{1}{2}\rho\,(1 - m^2) \int r^2\, d\omega \qquad \ldots\ldots\ldots\ldots\ldots(1)$$

integrated over the surface of the outer boundary.

From the equation of the ellipsoid in polar co-ordinates, we have

$$\frac{1}{r^2} = \sin^2\theta \left(\frac{\cos^2\phi}{a^2} + \frac{\sin^2\phi}{b^2} \right) + \frac{\cos^2\theta}{c^2},$$

and putting $d\omega = \sin\theta\, d\theta\, d\phi$, we have

$$\int r^2\, d\omega = 8 \int_0^{\frac{1}{2}\pi} \int_0^{\frac{1}{2}\pi} \frac{\sin\theta\, d\theta\, d\phi}{\sin^2\theta \left(\dfrac{\cos^2\phi}{a^2} + \dfrac{\sin^2\phi}{b^2} \right) + \dfrac{\cos^2\theta}{c^2}}$$

$$= 8 \int_0^{\frac{1}{2}\pi} \int_0^{\infty} \frac{\sin\theta\, d\theta\, dt}{\dfrac{\sin^2\theta}{a^2} + \dfrac{\cos^2\theta}{c^2} + \left(\dfrac{\sin^2\theta}{b^2} + \dfrac{\cos^2\theta}{c^2} \right) t^2},$$

where $t = \tan\phi$. But

$$\int_0^\infty \frac{dt}{A + Bt^2} = \left[\frac{1}{\sqrt{(AB)}}\tan^{-1}t\sqrt{\frac{B}{A}}\right]_0^\infty = \frac{\pi}{2\sqrt{(AB)}};$$

therefore

$$\int r^2 d\omega = 4\pi \int_0^{\frac12\pi} \frac{\sin\theta\, d\theta}{\sqrt{\left\{\left(\dfrac{\sin^2\theta}{a^2}+\dfrac{\cos^2\theta}{c^2}\right)\left(\dfrac{\sin^2\theta}{b^2}+\dfrac{\cos^2\theta}{c^2}\right)\right\}}}.$$

Put $u = c^2\tan^2\theta$, and we find that

$$\int r^2 d\omega = 2\pi abc \int_0^\infty \frac{du}{\sqrt{\{(a^2+u)(b^2+u)(c^2+u)\}}} \quad \dots(2).$$

We shall denote $\displaystyle\int_0^\infty \frac{du}{\sqrt{\{(a^2+u)(b^2+u)(c^2+u)\}}}$ by I, then substituting in (1), we find that the potential of a homoeoid at an internal point is

$$V = \pi\rho abc\,(1-m^2)\,I.$$

But the mass $M = \frac43\pi\rho abc\,(1-m^3)$, so that

$$V = \frac34 MI\frac{1-m^2}{1-m^3} \quad\dots\dots\dots\dots(3).$$

If the homoeoid be thin, then m is nearly equal to 1, say $1-\epsilon$, so that $\dfrac{1-m^2}{1-m^3} = \dfrac{2\epsilon}{3\epsilon} = \dfrac{2}{3}$, and

$$V = \tfrac12 MI \quad\dots\dots\dots\dots(4).$$

7·22. Attraction of a thin homoeoid at an external point. We shall prove first that this is directed along the normal to the confocal ellipsoid through the external point.

Let QP, QT be the normal and the tangent plane to the confocal at Q. Let the normal QP meet the plane of contact RR' of the enveloping cone of the homoeoid from Q in P. Through P as vertex take a cone of small solid angle $d\omega$ cutting the shell in elements MN, $M'N'$, and let MM' meet the tangent plane to the confocal at T. Then the pole with regard to the homoeoid of the tangent plane QT to the confocal lies on QP, but it also lies on the polar plane of Q, i.e. RR';

therefore P is the pole of the tangent plane QT. Therefore $(MPM'T)$ is a harmonic range. But PQT is a right angle, therefore QP bisects the angle MQM'.

Now if μ, μ' denote the masses of the elements MN, $M'N'$ of the shell, $\mu : \mu' = MP^2 : PM'^2$. But since QP bisects the angle MQM', therefore $MP : PM' = MQ : QM'$.

Therefore $\mu/MQ^2 = \mu'/M'Q^2$, or the attractions at Q of the elements MN, $M'N'$ are equal; and they are equally inclined to QP, therefore their resultant acts along QP. Since the same holds good for all like pairs of elements into which the homoeoid can be divided, it follows that the resultant attraction at Q is along QP.

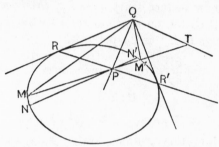

It is also evident that the plane of contact of the enveloping cone from Q divides the homoeoid into two parts which exert equal attractions at Q.

Since the resultant force at a point external to the thin homoeoid is always directed along the normal to the confocal ellipsoid through the point, it follows that the external equipotential surfaces are the confocal ellipsoids.

It follows that any two thin confocal homoeoids have the same external equipotential surfaces, viz. the ellipsoids confocal with them and external to both, and therefore by 4·31 (iii) their attractions at any point are proportional to their masses.

Now let a, b, c be the semi-axes of a thin homoeoid of mass M and density ρ, and let a', b', c' be the semi-axes of the confocal thin homoeoid having the point Q on its outer surface and of the same density. If p be the central perpendicular on the tangent plane at Q we may take dp as the thickness and $\rho\,dp$ as the mass per unit area of this confocal homoeoid. Since normal attraction increases by $4\pi\rho\,dp$ in crossing from inside to outside the surface (3·7) and the attraction is zero inside, therefore the attraction at Q of the outer homoeoid is $4\pi\rho\,dp$, but $dp/p = da'/a'$ (7·1), and the mass $M' = 4\pi\rho b'c'\,da'$ (7·1), so that the attraction of the outer homoeoid is $M'p/a'b'c'$.

But the attractions of the two homoeoids are proportional to their masses, therefore the attraction at Q of the original homoeoid is $Mp/a'b'c'$, where a', b', c' are semi-axes of the confocal through Q and p is the central perpendicular on the tangent plane to the confocal through Q.

We may now deduce *the potential of the homoeoid at an external point* Q, for since p is normal to the equipotential surface through Q, if V is the potential

$$\frac{\partial V}{\partial p} = \text{force in direction } p$$

$$= -Mp/a'b'c'$$

or $\qquad dV = -Mp\,dp/\sqrt{\{(a^2+\lambda)(b^2+\lambda)(c^2+\lambda)\}},$

where $\qquad x^2/(a^2+\lambda) + y^2/(b^2+\lambda) + z^2/(c^2+\lambda) = 1 \ldots\ldots\ldots(1)$

is the confocal through Q. But

$$p^2 = (a^2+\lambda)\,l^2 + (b^2+\lambda)\,m^2 + (c^2+\lambda)\,n^2$$

$$= a^2l^2 + b^2m^2 + c^2n^2 + \lambda,$$

so that $\qquad 2p\,dp = d\lambda$

and $\qquad dV = -\tfrac{1}{2}M\,d\lambda/\sqrt{\{(a^2+\lambda)(b^2+\lambda)(c^2+\lambda)\}}.$

Hence $\qquad V = \tfrac{1}{2}M\int_\lambda^\infty \dfrac{du}{\sqrt{\{(a^2+u)(b^2+u)(c^2+u)\}}} \quad\ldots\ldots(2),$

where the lower limit λ is given in terms of the co-ordinates x, y, z of Q, as the greatest root of the cubic in λ, (1), and the upper limit is determined by the consideration that $V \to 0$ as $\lambda \to \infty$.

7·3. Attraction of a homogeneous solid ellipsoid at an internal point. Let P be a point inside the ellipsoid. If we

draw a similar, similarly situated concentric ellipsoid through
P, the matter outside this surface constitutes a homoeoid and
exerts no attraction at P, so the
problem is reduced to finding the
attraction of an ellipsoid at a point
P on its surface.

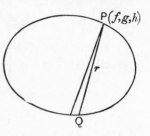

Let $x^2/a^2 + y^2/b^2 + z^2/c^2 = 1$ be the
ellipsoid through P and f, g, h the
co-ordinates of P. Let r be the length
of any chord PQ and l, m, n the
direction cosines of QP, so that the
co-ordinates of Q are $f - lr$, $g - mr$, $h - nr$. Then since Q and
P both lie on the ellipsoid, we have

$$r^2\left(\frac{l^2}{a^2} + \frac{m^2}{b^2} + \frac{n^2}{c^2}\right) - 2r\left(\frac{fl}{a^2} + \frac{gm}{b^2} + \frac{hn}{c^2}\right) = 0 \quad \ldots\ldots(1).$$

Taking a cone of small solid angle $d\omega$ with vertex at P and
axis along PQ, the attraction at P of the matter in this cone
is $\int \rho \, d\omega \, dr = \rho r \, d\omega$, where ρ is the density. Hence the components of the resultant attraction at P are given by

$$X = -\rho\int lr \, d\omega, \quad Y = -\rho\int mr \, d\omega, \quad Z = -\rho\int nr \, d\omega \ldots(2).$$

We must now substitute for r from (1) and integrate with
regard to $d\omega$ for cones in all directions on one side of the
tangent plane at P. But since every straight line through P
has a real intersection with the ellipsoid, if we integrate in all
directions round P we should take the ellipsoid twice over.

Therefore $\qquad X = -\tfrac{1}{2}\rho \int \dfrac{2l\left(\dfrac{fl}{a^2} + \dfrac{gm}{b^2} + \dfrac{hn}{c^2}\right)}{\dfrac{l^2}{a^2} + \dfrac{m^2}{b^2} + \dfrac{n^2}{c^2}} \, d\omega,$

where l, m, n may now be regarded as co-ordinates of a point
on a unit sphere, $d\omega$ an element of area of this sphere and
the integration extends to the whole surface. The terms in
lm and ln contribute nothing to the integral because terms lm
and $l(-m)$ occur equally and cancel one another.

Therefore $X = -\rho f \displaystyle\int \frac{\dfrac{l^2}{a^2}}{\dfrac{l^2}{a^2} + \dfrac{m^2}{b^2} + \dfrac{n^2}{c^2}} d\omega = -A\rho f$, say

Similarly $Y = -\rho g \displaystyle\int \frac{\dfrac{m^2}{b^2}}{\dfrac{l^2}{a^2} + \dfrac{m^2}{b^2} + \dfrac{n^2}{c^2}} d\omega = -B\rho g$ $\qquad$ ...(3).

and $\qquad Z = -\rho h \displaystyle\int \frac{\dfrac{n^2}{c^2}}{\dfrac{l^2}{a^2} + \dfrac{m^2}{b^2} + \dfrac{n^2}{c^2}} d\omega = -C\rho h$

We notice that A, B, C are functions of the ratios of the axes, so that it is immaterial whether a, b, c are semi-axes of the given ellipsoid or of the similar one through P.

7·31. To shew that

$$A = 2\pi abc \int_0^\infty \frac{du}{\sqrt{\{(a^2+u)^3 (b^2+u)(c^2+u)\}}}.$$

We have $\qquad A = \displaystyle\int \frac{\dfrac{l^2}{a^2}}{\dfrac{l^2}{a^2} + \dfrac{m^2}{b^2} + \dfrac{n^2}{c^2}} d\omega,$

and if we use polar co-ordinates this may be written

$$A = 8 \int_0^{\frac{1}{2}\pi} \int_0^{\frac{1}{2}\pi} \frac{\dfrac{\cos^2\theta}{a^2}}{\dfrac{\cos^2\theta}{a^2} + \sin^2\theta\left(\dfrac{\cos^2\phi}{b^2} + \dfrac{\sin^2\phi}{c^2}\right)} \sin\theta \, d\theta \, d\phi$$

$$= 8 \int_0^{\frac{1}{2}\pi} \int_0^\infty \frac{\dfrac{\cos^2\theta}{a^2} \sin\theta \, d\theta \, dt}{\dfrac{\cos^2\theta}{a^2} + \dfrac{\sin^2\theta}{b^2} + \left(\dfrac{\cos^2\theta}{a^2} + \dfrac{\sin^2\theta}{c^2}\right) t^2}, \text{ where } t = \tan\phi,$$

$$= 4\pi \int_0^{\frac{1}{2}\pi} \frac{\dfrac{\cos^2\theta}{a^2} \sin\theta \, d\theta}{\sqrt{\left\{\left(\dfrac{\cos^2\theta}{a^2} + \dfrac{\sin^2\theta}{b^2}\right)\left(\dfrac{\cos^2\theta}{a^2} + \dfrac{\sin^2\theta}{c^2}\right)\right\}}}.$$

Put $a^2 \tan^2 \theta = u$, and on reduction we get

$$A = 2\pi abc \int_0^\infty \frac{du}{\sqrt{\{(a^2+u)^3 (b^2+u)(c^2+u)\}}} \quad \ldots\ldots(1),$$

and similar expressions for B and C.

7·32. Spheroids. We see from **7·3** (3) that

$$\left. \begin{aligned} A+B+C &= \int d\omega = 4\pi \\ Aa^2 + Bb^2 + Cc^2 &= \int \frac{d\omega}{\dfrac{l^2}{a^2}+\dfrac{m^2}{b^2}+\dfrac{n^2}{c^2}} \end{aligned} \right\} \quad \ldots\ldots\ldots\ldots(1).$$

and

Two of the axes of a spheroid are equal, and if $a = b$ then

$$A = B, \quad X = -A\rho f, \quad Y = -A\rho g, \quad Z = -C\rho h,$$

and (1) become $\qquad 2A + C = 4\pi,$

$$\begin{aligned} 2Aa^2 + Cc^2 &= \int \frac{d\omega}{\dfrac{l^2+m^2}{a^2}+\dfrac{n^2}{c^2}} \\ &= \int_0^\pi \int_0^{2\pi} \frac{\sin\theta\, d\theta\, d\phi}{\dfrac{\sin^2\theta}{a^2}+\dfrac{\cos^2\theta}{c^2}} \\ &= 2\pi \int_{-1}^1 \frac{du}{\dfrac{1}{a^2}+\left(\dfrac{1}{c^2}-\dfrac{1}{a^2}\right)u^2}, \end{aligned}$$

where $u = \cos\theta$.

There are two cases:

(i) *Oblate spheroid*, $a > c$.

$$\begin{aligned} 2Aa^2 + Cc^2 &= \frac{4\pi a^2 c}{\sqrt{(a^2-c^2)}} \tan^{-1} \sqrt{\left(\frac{a^2}{c^2}-1\right)} \\ &= 4\pi a^2 \frac{\sqrt{(1-e^2)}}{e} \tan^{-1} \frac{e}{\sqrt{(1-e^2)}} \quad \ldots\ldots(2), \end{aligned}$$

where e is the eccentricity of the generating ellipse.

(ii) *Prolate spheroid*, $c > a$.

$$\begin{aligned} 2Aa^2 + Cc^2 &= \frac{2\pi a^2 c}{\sqrt{(c^2-a^2)}} \log \frac{c+\sqrt{(c^2-a^2)}}{c-\sqrt{(c^2-a^2)}} \\ &= 2\pi c^2 \frac{1-e^2}{e} \log \frac{1+e}{1-e} \quad \ldots\ldots\ldots\ldots(3), \end{aligned}$$

where $a^2 = c^2(1-e^2)$, so that e is the eccentricity of the generating ellipse.

A and C can then be found in either case, since $2A + C = 4\pi$.

7·33. Attraction of an oblate spheroid of small ellipticity.
When the ellipticity ϵ is small, since $c=a(1-\epsilon)$ and $c^2=a^2(1-e^2)$,
therefore $e^2=2\epsilon$ approximately.

Hence from 7·32 (i),

$$2A+C(1-e^2)=4\pi\,\frac{\sqrt{(1-e^2)}}{e}\tan^{-1}\frac{e}{\sqrt{(1-e^2)}}$$

$$=4\pi\left\{1-\frac{1}{3}\frac{e^2}{1-e^2}+\frac{1}{5}\frac{e^4}{(1-e^2)^2}-\cdots\right\}.$$

Whence, since $2A+C=4\pi$,

$$Ce^2=4\pi\left\{\frac{1}{3}\frac{e^2}{1-e^2}-\frac{1}{5}\frac{e^4}{(1-e^2)^2}\right\}$$

or

$$C=\frac{4\pi}{3}\left\{1+e^2+\cdots-\frac{3e^2}{5}-\cdots\right\}$$

$$=\tfrac{4}{3}\pi(1+\tfrac{1}{5}\epsilon).$$

But $2A+C=4\pi$, so that $A=\tfrac{4}{3}\pi(1-\tfrac{2}{5}\epsilon)$.

Hence in this case

$$\left.\begin{aligned}X&=-\tfrac{4}{3}\pi\rho(1-\tfrac{2}{5}\epsilon)f\\ Y&=-\tfrac{4}{3}\pi\rho(1-\tfrac{2}{5}\epsilon)g\\ Z&=-\tfrac{4}{3}\pi\rho(1+\tfrac{4}{5}\epsilon)h\end{aligned}\right\}\quad\dots\dots\dots\dots\dots(1).$$

These results are of importance in connection with figures of equilibrium of rotating liquid.

7·34. Attractions of a nearly spherical ellipsoid. Let
$b=a(1-\eta)$ and $c=a(1-\epsilon)$. Neglecting higher powers of η and ϵ
than the first, we may put $A=A_0+F\eta+F'\epsilon$.

If the ellipsoid were turned through a right angle about Ox, b and c
would be interchanged, so we must have $F=F'$. Also an oblate
spheroid for which $b=a$ and $\eta=0$ is a special case, for which, from 7·33,

$$A=\tfrac{4}{3}\pi(1-\tfrac{2}{5}\epsilon),$$

therefore

$$A_0=\tfrac{4}{3}\pi\quad\text{and}\quad F'=-\tfrac{8}{15}\pi.$$

Hence

$$A=\tfrac{4}{3}\pi(1-\tfrac{2}{5}\eta-\tfrac{2}{5}\epsilon).$$

Similarly

$$B=\tfrac{4}{3}\pi\left\{1-\frac{2}{5}\left(1-\frac{a}{b}\right)-\frac{2}{5}\left(1-\frac{c}{b}\right)\right\}$$

$$=\tfrac{4}{3}\pi\{1+\tfrac{4}{5}\eta-\tfrac{2}{5}(\epsilon-\eta)\}$$

$$=\tfrac{4}{3}\pi(1+\tfrac{6}{5}\eta-\tfrac{2}{5}\epsilon)$$

and

$$C=\tfrac{4}{3}\pi(1-\tfrac{2}{5}\eta+\tfrac{4}{5}\epsilon).$$

Therefore

$$\left.\begin{aligned}X&=-\tfrac{4}{3}\pi\rho(1-\tfrac{2}{5}\eta-\tfrac{2}{5}\epsilon)f\\ Y&=-\tfrac{4}{3}\pi\rho(1+\tfrac{6}{5}\eta-\tfrac{2}{5}\epsilon)g\\ Z&=-\tfrac{4}{3}\pi\rho(1-\tfrac{2}{5}\eta+\tfrac{4}{5}\epsilon)h\end{aligned}\right\}\quad\dots\dots\dots\dots\dots(1).$$

We may obtain similar results in a symmetrical form, if we take k as the mean radius $\frac{1}{3}(a+b+c)$, then

$$A = \tfrac{4}{3}\pi\left(1 - \frac{2}{5}\frac{2a-b-c}{a}\right)$$

$$= \tfrac{4}{3}\pi\left(1 - \frac{6}{5}\frac{a-k}{k}\right)$$

Similarly, $\qquad B = \tfrac{4}{3}\pi\left(1 - \frac{6}{5}\frac{b-k}{k}\right)$ ....(2).

and $\qquad C = \tfrac{4}{3}\pi\left(1 - \frac{6}{5}\frac{c-k}{k}\right)$

7·35. Potential of a homogeneous solid ellipsoid at an internal point. From **7·21** (3) it follows that, since a solid ellipsoid can be divided into similar homoeoids, the innermost being of zero dimensions, by putting $m=0$, the potential *at the centre* of a homogeneous solid ellipsoid is

$$V_0 = \tfrac{3}{4}MI = \pi\rho abc \int_0^\infty \frac{du}{\sqrt{\{(a^2+u)(b^2+u)(c^2+u)\}}} \quad ...(1).$$

To find the potential at an internal point (x, y, z), we make use of the fact that the attraction components are given by

$$X, Y, Z = -A\rho x, \ -B\rho y, \ -C\rho z \quad (7\cdot3) \ \ ......(2).$$

Hence, if V denotes the potential,

$$dV = \frac{\partial V}{\partial x}dx + \frac{\partial V}{\partial y}dy + \frac{\partial V}{\partial z}dz$$

$$= X\,dx + Y\,dy + Z\,dz$$

$$= -\rho\,(Ax\,dx + By\,dy + Cz\,dz).$$

Therefore $\qquad V = \tfrac{1}{2}\rho\,(D - Ax^2 - By^2 - Cz^2) \quad ....(3),$

where the constant of integration $\tfrac{1}{2}\rho D$ must be the potential at the centre. Hence by substituting for D from (1) above, and for A, B, C from **7·31**, we get

$$V = \pi\rho abc \int_0^\infty \frac{1}{\sqrt{\{(a^2+u)(b^2+u)(c^2+u)\}}}$$

$$\times \left(1 - \frac{x^2}{a^2+u} - \frac{y^2}{b^2+u} - \frac{z^2}{c^2+u}\right)du \ \ ......(4).$$

It follows from (3) that the equipotential surfaces inside the ellipsoid are ellipsoids

$$Ax^2 + By^2 + Cz^2 = \text{const.} \quad \ldots\ldots\ldots\ldots(5),$$

similar to one another though not similar to the boundary of the solid.

7·4. Ivory's Theorem.* Attraction of a homogeneous solid ellipsoid at an external point.

If x, y, z and x', y', z' are the co-ordinates of points P and P' one on each of two confocal ellipsoids

$$x^2/a^2 + y^2/b^2 + z^2/c^2 = 1 \quad \ldots\ldots\ldots\ldots\ldots(1),$$

and

$$x^2/a'^2 + y^2/b'^2 + z^2/c'^2 = 1 \quad \ldots\ldots\ldots\ldots(2),$$

then P and P' are called *corresponding points* if

$$x/a = x'/a', \quad y/b = y'/b', \quad z/c = z'/c' \quad \ldots\ldots\ldots(3).$$

Let RS be an elementary strip of the first ellipsoid of cross-section $dy\,dz$, parallel to the x-axis; and let $R'S'$ be the corresponding strip of the second ellipsoid of cross-section $dy'\,dz'$, so that

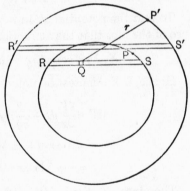

$$dy\,dz/dy'\,dz' = bc/b'c'.$$

Then if $f'(r)$ denotes the law of force at distance r, and ρ the density of either ellipsoid, the component parallel to the x-axis of the attraction at P' due to the strip RS

$$= -\rho\,dy\,dz \int f'(r) \cos P'QS\,dx,$$

when Q is the position of an element $dx\,dy\,dz$. This

$$= -\rho\,dy\,dz \int f'(r)\left(-\frac{dr}{dx}\right)dx = \rho\,dy\,dz\{f(P'S) - f(P'R)\}.$$

* Sir James Ivory (1765–1842). Scottish mathematician.

In the same way the x-component of the attraction at P due to the strip $R'S'$

$$= \rho\, dy'\, dz'\, \{f(PS') - f(PR')\}.$$

But it is easy to deduce from (3) that for any pairs of corresponding points $PS' = P'S$ and $PR' = P'R$.

Hence the ratio of the x-component of the attraction of RS at P' to the similar component of $R'S'$ at P is

$$\frac{dy\, dz}{dy'\, dz'} = \frac{bc}{b'c'} \qquad \dots\dots\dots\dots\dots(4).$$

If in like manner we take all such strips as RS of the first ellipsoid and all such strips as $R'S'$ of the second, we get

$$X : X' = bc : b'c' \qquad \dots\dots\dots\dots\dots(5),$$

where X is the x-component of the attraction of the first ellipsoid at P' and X' is the x-component of the attraction of the second ellipsoid at P.

This result is **Ivory's theorem** and is true for any law of force.

Now, from **7·3**, $X' = -A'\rho x$, where P is the point (x, y, z) and A' is the same function of a', b', c' that A is of a, b, c.

Therefore

$$X = -\frac{bc}{b'c'} A'\rho x = -\frac{abc}{b'c'} A'\rho \frac{x}{a} = -\frac{abc}{a'b'c'} A'\rho x', \quad \text{from (3)},$$

where x', y', z' are the co-ordinates of P'.

Similar relations hold for the y and z components of attraction, so that the attraction at P', (x', y', z'), of the ellipsoid (a, b, c) has components

$$\left. \begin{aligned} X &= -\frac{abc}{a'b'c'} A'\rho x' \\[2mm] Y &= -\frac{abc}{a'b'c'} B'\rho y' \\[2mm] Z &= -\frac{abc}{a'b'c'} C'\rho z' \end{aligned} \right\} \qquad \dots\dots\dots\dots(6),$$

where a', b', c' are the semi-axes of the confocal ellipsoid which passes through P', and A', B', C' are the same functions of a', b', c' that A, B, C are of a, b, c.

7·41. MacLaurin's Theorem.* *The attractions of two confocal ellipsoids at a point external to both are proportional to their masses and in the same direction.*

In 7·22 this was proved to be true of confocal thin homoeoids; that it is true for homogeneous solid confocal ellipsoids follows as a direct corollary from 7·4 (6), if we regard P' as a fixed point and consider the attractions at P' of two different confocal ellipsoids, of semi-axes a, b, c and density ρ and of semi-axes a_1, b_1, c_1 and density ρ_1; the x, y, and z components are all in the ratio $\rho abc : \rho_1 a_1 b_1 c_1$ and are therefore proportional to the masses and have resultants in the same direction.

7·42. Potential of a solid homogeneous ellipsoid at an external point. Let a, b, c be the semi-axes of the given ellipsoid and x, y, z the co-ordinates of the external point P. Let a', b', c' be the semi-axes of a confocal ellipsoid which passes through P, and V, V' the potentials at P of the first and second ellipsoids regarded as of the same density.

Then by MacLaurin's theorem (7·41)

$$V : V' = abc : a'b'c'.$$

But, from 7·35 (4),

$$V' = \pi\rho a'b'c' \int_0^\infty \frac{1}{\sqrt{\{(a'^2+u)(b'^2+u)(c'^2+u)\}}}$$
$$\times \left(1 - \frac{x^2}{a'^2+u} - \frac{y^2}{b'^2+u} - \frac{z^2}{c'^2+u}\right) du.$$

Therefore

$$V = \pi\rho abc \int_0^\infty \frac{1}{\sqrt{\{(a'^2+u)(b'^2+u)(c'^2+u)\}}}$$
$$\times \left(1 - \frac{x^2}{a'^2+u} - \frac{y^2}{b'^2+u} - \frac{z^2}{c'^2+u}\right) du \quad \ldots(1);$$

but since the ellipsoids are confocal, we have

$$a'^2, b'^2, c'^2 = a^2+\lambda, b^2+\lambda, c^2+\lambda,$$

where λ is the largest root of the cubic

$$\frac{x^2}{a^2+\lambda} + \frac{y^2}{b^2+\lambda} + \frac{z^2}{c^2+\lambda} = 1 \quad \ldots\ldots\ldots\ldots(2).$$

* Colin MacLaurin (1698–1746). Scottish mathematician.

Substituting in (1) for a'^2, b'^2, c'^2, and writing v for $\lambda+u$, we get

$$V = \pi\rho abc \int_\lambda^\infty \frac{1}{\sqrt{\{(a^2+v)(b^2+v)(c^2+v)\}}}$$
$$\times \left(1 - \frac{x^2}{a^2+v} - \frac{y^2}{b^2+v} - \frac{z^2}{c^2+v}\right) dv \quad \dots(3),$$

where the lower limit λ is determined by (2) as stated.

7·5. Attraction of an infinitely long homogeneous solid elliptic cylinder. This can be deduced from **7·3** and **7·4** by making $c \to \infty$.

(i) *Internal point.* The components of attraction at a point (x, y, z) are given by

$$X = -A\rho x, \quad Y = -B\rho y, \quad Z = 0;$$

where, from **7·31**,

$$A = 2\pi ab \int_0^\infty \frac{du}{\sqrt{\{(a^2+u)^3(b^2+u)\}}} \cdot$$

By substituting $\dfrac{1}{v^2}$ for a^2+u, we get

$$A = 4\pi ab \int_0^{1/a} \frac{v\,dv}{\sqrt{\{1-v^2(a^2-b^2)\}}}$$

or

$$A = \frac{4\pi b}{a+b} \cdot$$

Similarly

$$B = \frac{4\pi a}{a+b};$$

and

$$X = -\frac{4\pi\rho ab}{a+b} \cdot \frac{x}{a}; \quad Y = -\frac{4\pi\rho ab}{a+b} \cdot \frac{y}{b} \quad \dots\dots\dots(1).$$

(ii) *External point.* In like manner from **7·4** we find for the attraction components at an external point (x, y)

$$X = -\frac{4\pi\rho ab}{a'+b'} \cdot \frac{x}{a'}; \quad Y = -\frac{4\pi\rho ab}{a'+b'} \cdot \frac{y}{b'} \quad \dots\dots\dots(2),$$

where a', b' are the semi-axes of an ellipse confocal with the cross-section of the cylinder and passing through (x, y).

7·6. Equilibrium of rotating liquid. When a mass of gravitating liquid rotates uniformly about an axis, it is conceivable that for a certain form of the free surface the liquid particles may be in relative equilibrium. Since the resultant attraction on a particle depends on the form of the boundary surface, which is unknown, the problem does not admit of a complete solution. But assuming the liquid to be homogeneous, it can be shewn that certain forms are possible forms of relative equilibrium.

Thus if ρ be the density and ω the angular velocity, whether spheroidal or ellipsoidal forms are possible forms of relative equilibrium is found to depend on the numerical value of $\omega^2/2\pi\rho$.* We shall illustrate the theory by considering two simple cases.

7·61. *To shew that an* **oblate spheroid of small ellipticity** *is a possible form of relative equilibrium of a mass of uniform liquid rotating about an axis with a small uniform angular velocity.* (**MacLaurin's theorem.**)

The problem can be treated as a statical one if we compound with the force of attraction the reversed effective force, $\omega^2 r$ per unit mass, where r denotes distance from the axis of rotation.

The equation for the pressure is then

$$\frac{dp}{\rho} = X\,dx + Y\,dy + Z\,dz + \omega^2\,(x\,dx + y\,dy)$$

or

$$\frac{dp}{\rho} = -A\rho\,(x\,dx + y\,dy) - C\rho z\,dz + \omega^2\,(x\,dx + y\,dy),$$

assuming that the boundary is an oblate spheroid.

But the pressure must be constant over the boundary, so that its equation must be of the form

$$(\omega^2 - A\rho)\,(x^2 + y^2) - C\rho z^2 = \text{const.} \quad \ldots\ldots\ldots(1).$$

If $a, a, a\,(1 - \epsilon)$ are the semi-axes of the oblate spheroid and the ellipticity ϵ is small, its equation is

$$x^2 + y^2 + \frac{z^2}{1 - 2\epsilon} = a^2 \quad \ldots\ldots\ldots\ldots(2).$$

---

* Besant and Ramsey, *Treatise on Hydromechanics*, Part I, ch. VIII, where a discussion of the subject with full references will be found.

By comparing (1) and (2), we get

$$\omega^2 - A\rho = -C\rho\,(1 - 2\epsilon) \quad\ldots\ldots\ldots\ldots(3).$$

But for an oblate spheroid of small ellipticity, from **7·33**,

$$A = \tfrac{4}{3}\pi\,(1 - \tfrac{2}{5}\epsilon) \quad\text{and}\quad C = \tfrac{4}{3}\pi\,(1 + \tfrac{4}{5}\epsilon);$$

and, by substituting these values in (3), we find that $\epsilon = \dfrac{15\,\omega^2}{16\,\pi\rho}$,

and with this relation between ϵ and ω the condition is satisfied.

7·62. Jacobi's ellipsoid.* *An ellipsoid with three unequal axes is also a possible form.*

For an ellipsoid the pressure equation is

$$\frac{dp}{\rho} = -A\rho x\,dx - B\rho y\,dy - C\rho z\,dz + \omega^2\,(x\,dx + y\,dy),$$

where A, B, C are given by **7·31**.

The surfaces of constant pressure are therefore

$$(\omega^2 - A\rho)\,x^2 + (\omega^2 - B\rho)\,y^2 - C\rho z^2 = \text{const.}\ \ldots\ldots(1).$$

And if we assume the free surface to be

$$\frac{x^2}{a^2} + \frac{y^2}{b^2} + \frac{z^2}{c^2} = 1 \quad\ldots\ldots\ldots\ldots\ldots(2),$$

we get by comparing (1) and (2)

$$a^2\,(\omega^2 - A\rho) = b^2\,(\omega^2 - B\rho) = -c^2 C\rho \ \ldots\ldots(3).$$

By eliminating ω, we get

$$a^2 b^2\,(B - A) = (a^2 - b^2)\,c^2 C \ \ldots\ldots\ldots\ldots(4).$$

If we put $D = \{(a^2 + u)\,(b^2 + u)\,(c^2 + u)\}^{\frac{1}{2}}$, then, from **7·31**,

$$A = 2\pi abc \int_0^\infty \frac{du}{(a^2 + u)\,D}, \quad B = 2\pi abc \int_0^\infty \frac{du}{(b^2 + u)\,D},$$

$$C = 2\pi abc \int_0^\infty \frac{du}{(c^2 + u)\,D};$$

and (4) becomes

$$(a^2 - b^2) \int_0^\infty \frac{du}{D} \left\{ \frac{a^2 b^2}{(a^2 + u)\,(b^2 + u)} - \frac{c^2}{c^2 + u} \right\} = 0.$$

* Carl Gustav Jacob Jacobi (1804–1851). German mathematician.

Hence, if $a \neq b$, the lengths of the axes must be such as to satisfy the equation

$$\int_0^\infty \frac{u\,du}{D^3}\left(\frac{1}{a^2}+\frac{1}{b^2}-\frac{1}{c^2}+\frac{u}{a^2b^2}\right)=0 \quad\ldots\ldots\ldots\ldots(5).$$

If a and b are given, then (5) is an equation for determining c, and since the left-hand member is negative when c is small, and becomes positive as c increases, there must be a real value of c which satisfies the equation.

Further, ω^2 is given from (3) in the form

$$(a^2-b^2)\,\omega^2 = (Aa^2 - Bb^2)\rho$$

$$= \tfrac{3}{2}M\,(a^2-b^2)\int_0^\infty \frac{u\,du}{(a^2+u)\,(b^2+u)\,D},$$

where M is the mass of liquid. So that when $a \neq b$,

$$\omega^2 = \tfrac{3}{2}M\int_0^\infty \frac{u\,du}{(a^2+u)\,(b^2+u)\,D};$$

and since this expression is positive there is a real value of ω, thus establishing that the ellipsoid is a possible form.

Further, from (3),

$$a^2\omega^2 = (Aa^2 - Cc^2)\rho$$

$$= \tfrac{3}{2}M\,(a^2-c^2)\int_0^\infty \frac{u\,du}{(a^2+u)\,(c^2+u)\,D},$$

so that for ω to be real we must have $c < a$. Similarly $c < b$, so that the ellipsoid must rotate about its least axis.

7·63. Example. *A rigid gravitating sphere of radius a and uniform density ρ is surrounded by a layer of gravitating liquid of volume $\tfrac{4}{3}\pi\,(b^3 - a^3)$ and uniform density σ and placed midway between two spheres of mass M whose distance apart $(2f)$ is so great that powers of b/f above the third may be neglected. Prove that the boundary of the liquid is a prolate spheroid of small ellipticity*

$$\frac{45Mb^3}{4\pi f^3\{5a^3\,(\rho - \sigma) + 2\sigma b^3\}},$$

and determine the small angular velocity of the whole mass which, to this approximation, would make the boundary a sphere.   [P. 1933]

Let A, B be the centres of the two spheres of mass M and O that of the sphere of radius a. Let Q be a point in the liquid at distance r from O, where the angle $QOB = \theta$, and let OB be the axis of z.

It is easy to shew that the potential at Q due to the two spheres of mass M, being $\dfrac{M}{AQ} + \dfrac{M}{BQ}$, is equal to $\dfrac{2M}{f}\left(1 + \dfrac{r^2}{f^2}P_2\right)$, neglecting higher powers than b^3/f^3. This represents a small disturbing field of force which distorts the boundary of the fluid from what would otherwise be a sphere. We assume therefore that the boundary surface is $r = b(1 + \epsilon P_2)$, where ϵ is small, introducing the same Legendre coefficient as occurs in the potential of the disturbing field.

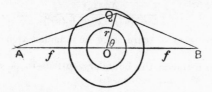

We now regard the potential at Q as due to the spheres at A and B, a solid sphere of radius a and density $\rho - \sigma$, on which is superposed another solid sphere of radius b and density σ, together with a surface distribution of density $\sigma b \epsilon P_2$.

If we assume that this layer produces potentials

$$V_1 = A\frac{r^2}{b^2}P_2 \quad (r < b) \quad \text{and} \quad V_2 = A\frac{b^3}{r^3}P_2 \quad (r > b),$$

we have
$$4\pi\sigma b\epsilon P_2 = \left(\frac{\partial V_1}{\partial r} - \frac{\partial V_2}{\partial r}\right)_{r=b} = \frac{5AP_2}{b},$$

so that $A = \tfrac{4}{5}\pi\epsilon\sigma b^2$, and $V_1 = \tfrac{4}{5}\pi\epsilon\sigma r^2 P_2$.

The whole potential at Q is therefore

$$V = \tfrac{4}{3}\pi(\rho - \sigma)\frac{a^3}{r} + \tfrac{2}{3}\pi\sigma(3b^2 - r^2) + \tfrac{4}{5}\pi\epsilon\sigma r^2 P_2 + \frac{2M}{f}\left(1 + \frac{r^2}{f^2}P_2\right) \quad ......(1).$$

Let us also suppose that the whole has a slow angular velocity ω about AB, then the equation for the pressure in the liquid is

$$\frac{dp}{\rho} = dV + \omega^2(x\,dx + y\,dy),$$

so that
$$V + \tfrac{1}{2}\omega^2(x^2 + y^2) = \text{const.}$$

represents surfaces of constant pressure, and must include the boundary surface.

But $x^2 + y^2 = r^2\sin^2\theta = \tfrac{2}{3}r^2(1 - P_2)$, so that we must have

$$\tfrac{4}{3}\pi(\rho - \sigma)\frac{a^3}{r} + \tfrac{2}{3}\pi\sigma(3b^2 - r^2) + \tfrac{4}{5}\pi\epsilon\sigma r^2 P_2$$
$$+ \frac{2M}{f}\left(1 + \frac{r^2}{f^2}P_2\right) + \tfrac{1}{3}\omega^2 r^2(1 - P_2) = \text{const.}$$

at all points on the surface $r = b(1 + \epsilon P_2)$.

A comparison of the equation of the boundary surface with the expression for the disturbing field of force suggests that ϵ is of order $(b/f)^3$. Hence, by substituting for r, we find that the coefficient of P_2 vanishes if ω^2 is also of order $(b/f)^3$ and

$$-\tfrac{4}{3}\pi(\rho-\sigma)\frac{a^3}{b}\epsilon - \tfrac{4}{3}\pi\sigma b^2\epsilon + \tfrac{4}{5}\pi\sigma b^2\epsilon + 2M\frac{b^2}{f^3} - \tfrac{1}{3}\omega^2 b^2 = 0.$$

Hence, when there is no rotation, we have

$$\epsilon = \frac{15Mb^3}{2\pi f^3\{5(\rho-\sigma)a^3 + 2\sigma b^3\}} \quad\ldots\ldots\ldots\ldots\ldots(2),$$

and ϵ would be zero if there were an angular velocity given by $\omega^2 = 6M/f^3$.

But we have assumed the boundary surface to be $r = b(1+\epsilon P_2)$, or $r^2(1-2\epsilon P_2) = b^2$, which reduces to

$$\frac{x^2+y^2}{(1-\tfrac{1}{2}\epsilon)^2} + \frac{z^2}{(1+\epsilon)^2} = b^2,$$

and represents a prolate spheroid of ellipticity $\tfrac{3}{2}\epsilon$, and hence the required result.

EXAMPLES

1. Prove the formulae

$$A = \int\frac{x^2}{a^2}d\omega, \quad B = \int\frac{y^2}{b^2}d\omega, \quad C = \int\frac{z^2}{c^2}d\omega,$$

where the integrations are over the whole surface of the ellipsoid.

2. Shew that, if $a>b>c$, then $Aa^2 > Bb^2 > Cc^2$; and that at points on the surface of such an ellipsoid the potential is greatest at the end of the axis c. [London Univ. 1931]

3. The bounding surfaces of a thin homoeoid are prolate spheroids and a cone is drawn having its vertex on a directrix plane. Shew that the portions of the homoeoid cut out by this cone exert equal attractions at the focus which corresponds to the directrix plane. [M. T. 1898]

4. Prove that, if an ellipsoid is a level surface for any distribution of matter within it, the external level surfaces are confocal ellipsoids.

5. Find the distribution of matter which will produce zero potential outside the surface $x^2/a^2 + y^2/b^2 + z^2/c^2 = 1$ and potential

$$\tfrac{1}{2}(1 - x^2/a^2 - y^2/b^2 - z^2/c^2)$$

throughout the space enclosed by it.

Deduce MacLaurin's theorem on the attractions of confocal ellipsoids.

6. Prove that for a prolate spheroid, $[a = b = c(1-e^2)^{\frac{1}{2}}]$,

$$C = 4\pi \left\{ \tfrac{1}{3} - 2 \sum_{n=1}^{\infty} \frac{e^{2n}}{(2n+1)(2n+3)} \right\}.$$

[London Univ. 1925]

7. Prove that a spheroid of uniform density cannot have its boundary surface as one of its level surfaces.

8. Prove that the level surfaces inside a solid homogeneous ellipsoid are similar and similarly situated ellipsoids.

If from this ellipsoid a co-axial ellipsoid be removed, shew that the level surfaces within the cavity are hyperboloids whose asymptotic cones have each three mutually perpendicular generators.   [C. 1907]

9. Find the amount by which the gravitational potential energy of a uniform solid ellipsoid exceeds that of a uniform sphere of equal volume and mass.   [M. T. 1912]

10. A solid homogeneous ellipsoid is divided by a plane perpendicular to an axis. Prove that the mutual attraction of the parts for varying positions of the plane varies as the square of the area of the section.   [C. 1881]

11. Prove that the resultant attraction between the octant of the ellipsoid $x^2/a^2 + y^2/b^2 + z^2/c^2 = 1$ in which x, y, z are all positive and the rest of the body is a single force of magnitude

$$\tfrac{1}{16}\pi\rho^2 abc\,(A^2a^2 + B^2b^2 + C^2c^2)^{\frac{1}{2}}$$

acting along the line

$$\frac{15\pi x - 16a}{Aa} = \frac{15\pi y - 16b}{Bb} = \frac{15\pi z - 16c}{Cc},$$

where ρ is the density assumed to be uniform.

12. A uniform oblate spheroid is divided into two parts by its equatorial plane; shew that, at a point of the equator, the component attraction of either half perpendicular to the equatorial plane is

$$\frac{4\gamma\rho a\,(1-e^2)}{e^3}\,(\tanh^{-1}e - e),$$

where a is the equatorial radius and e the eccentricity of a principal elliptic section.   [London Univ.]

13. Shew that the resultant attraction of one half of a solid homogeneous oblate spheroid, cut off by an equatorial plane, at a point on the rim of the base is inclined to the plane of the base at an angle whose tangent is

$$4c\,(\tanh^{-1}e - e)/(a\sin^{-1}e - ce)\,\pi,$$

where a and c are the semi-axes and e the eccentricity of the meridian section.   [C. 1903]

14. Shew that for a nearly spherical ellipsoid for which

$$a = k(1+\lambda), \quad b = k(1+\mu), \quad c = k(1+\nu),$$

where λ, μ, ν are small, $k^3 = abc$ and $\lambda + \mu + \nu = 0$ approximately, the components of attraction at an internal point are

$$-\tfrac{4}{3}\pi\rho\left(1 - \tfrac{6}{5}\lambda\right)x, \quad -\tfrac{4}{3}\pi\rho\left(1 - \tfrac{6}{5}\mu\right)y, \quad -\tfrac{4}{3}\pi\rho\left(1 - \tfrac{6}{5}\nu\right)z.$$

15. Prove that, if a solid uniform ellipsoid of mass M is nearly spherical and has semi-axes μ, $(\mu^2 - h)^{\frac{1}{2}}$, $(\mu^2 - k)^{\frac{1}{2}}$, the potential at an external point is

$$\frac{M}{r} + \frac{1}{10}\frac{M}{r^5}\left\{x^2(h+k) + y^2(-2h+k) + z^2(-2k+h)\right\}$$

to the first order of small quantities. [M. T. 1895]

---

16. An ellipsoid is cut into two equal portions, by a plane whose equation referred to the principal axes is $lx + my + nz = 0$. Shew that the attraction of one half of the ellipsoid on the other reduces to a single force, whose x component is $\tfrac{1}{2}\pi\rho Aa^3bcl/p$, where $p^2 = a^2l^2 + b^2m^2 + c^2n^2$. [M. T. 1899]

17. A solid ellipsoid of uniform density ρ is bounded by

$$x^2/a^2 + y^2/b^2 + z^2/c^2 = 1,$$

and is divided by the plane $y = x\tan\alpha$. Prove that the force parallel to this plane on either part required to prevent it from sliding over the other part is

$$\tfrac{1}{2}\pi^2\rho^2 a^2 b^2 c^2 \frac{(a^2-b^2)\sin\alpha\cos\alpha}{(a^2\sin^2\alpha + b^2\cos^2\alpha)^{\frac{1}{2}}} \int_0^\infty \frac{u\,du}{\{(a^2+u)^3(b^2+u)^3(c^2+u)\}^{\frac{1}{2}}}.$$

18. Prove that the potential of an ellipsoid of uniform density ρ at an internal point (x, y, z) exceeds the potential of a thin shell of the same mass M coincident with its surface and bounded by confocal ellipsoids by

$$2\pi\rho\left(1 - \frac{x^2}{a^2} - \frac{y^2}{b^2} - \frac{z^2}{c^2}\right)\Big/\left(\frac{1}{a^2} + \frac{1}{b^2} + \frac{1}{c^2}\right),$$

where a, b, c are the semi-axes of the ellipsoid and its principal axes are the axes of co-ordinates. [M. T. 1894]

19. Prove that, if a sector of a homogeneous solid elliptic cylinder of infinite length is cut out by planes through the axis and through conjugate radii CP, CP' of the cross-section, then the resultant force on this sector is constant for all positions of the cutting planes, and its direction bisects the angle QCQ', where CQ, CQ' are the corresponding radii of the auxiliary circle of the cross-section. [M. T. 1904]

20. Prove by direct integration or otherwise that the components of attraction at an internal point (x, y, z) of a solid homogeneous cylinder of density ρ, whose cross-section is the ellipse $x^2/a^2 + y^2/b^2 = 1$, and whose length is infinite in both directions, are

$$-4\pi\gamma\rho\,\frac{ab}{a+b}\,\frac{x}{a}, \quad -4\pi\gamma\rho\,\frac{ab}{a+b}\,\frac{y}{b}, \quad 0.$$

If (x, y, z) is an external point, and a', b' are the semi-axes of the cylinder through the point which is confocal with the given cylinder, verify that, if X', Y', denote respectively

$$-4\pi\gamma\rho\,\frac{ab}{a'+b'}\,\frac{x}{a'}, \quad -4\pi\gamma\rho\,\frac{ab}{a'+b'}\,\frac{y}{b'},$$

then
$$\frac{\partial X'}{\partial y} = \frac{\partial Y'}{\partial x} \quad \text{and} \quad \frac{\partial X'}{\partial x} + \frac{\partial Y'}{\partial y} = 0.$$

Hence, or otherwise, shew that X', Y' are the components of attraction at the external point. [London Univ.]

21. Prove that, if the surface density of an elliptic disc of semi-axes a, b is inversely proportional to $(1 - x^2/a^2 - y^2/b^2)^{\frac{1}{2}}$, it has the same attraction at points external to both, as a thin homoeoid whose focal ellipse is the boundary of the disc.

22. Shew that any plane divides a homogeneous solid ellipsoid into two parts such that the attraction between them reduces to a single force. [C. 1891]

23. The space between a solid homogeneous ellipsoid and a concentric ellipsoidal envelope is filled with a homogeneous liquid; prove that the resultant attraction between the solid and liquid reduces to a couple, which vanishes only when the ellipsoids are co-axial.

[C. 1904]

24. Shew, by the method of 7·63, that an oblate spheroid of small ellipticity is a possible form of relative equilibrium of a mass of homogeneous gravitating liquid rotating slowly.

25. A particle of mass M is held at a fixed distance $r\,(>a)$ from the centre C of a uniform circular disc, and the line joining the particle to C makes an angle θ with the axis of the disc. Shew that, neglecting terms containing fifth or higher power of (a/r), the couple about C tending to increase θ is given by

$$\frac{3MG\pi a^4\rho}{4r^3}\cos\theta\sin\theta,$$

where ρ and a are the surface density and radius of the disc and G is the constant of gravitation. [P. 1935]

26. If the law of attraction were $\gamma mm'r$, shew that a finite mass of fluid, which is rotating with uniform angular velocity ω about a fixed axis, could only be in relative equilibrium if ω^2 were less than γM, M being the total mass of the fluid. Shew that the excess of the pressure at the origin over that at the surface is for a given mass of fluid proportional to $(\gamma M - \omega^2)^{\frac{3}{2}}$. [C. 1915]

27. A uniform gravitating liquid sphere of radius a and density ρ is made to rotate with small uniform angular velocity ω, its surface being free. Find the form of the free surface, and shew that the pressure at the point $(\frac{1}{2}a, \theta)$ is

$$\tfrac{1}{3}\pi G\rho^2 a^2 - \tfrac{7}{48}\rho\omega^2 a^2 - \tfrac{5}{16}\rho\omega^2 a^2 \cos^2\theta,$$

where G denotes the constant of gravitation. [M. T. 1928]

28. A homogeneous gravitating solid is in the form of a prolate spheroid of small ellipticity ϵ, where $\epsilon = (a-c)/a$, and a, c are the semi-axes of a meridian section. The spheroid is rotating about its axis with angular velocity ω. Prove that the direction of a plumb line at any point on the spheroid will pass through the centre provided $\omega^2 = 6\epsilon g/5a$, where g is the approximately constant value of gravity at the surface. [M. T. 1927]

29. A planet is at the centre of a uniform plane annulus of mass M and inner and outer radii a and b, large compared with the planet's radius. Shew that to a first approximation the effect of the ring on the figure of equilibrium of the planet is equivalent to that of a rotational velocity ω, where $\omega^2 = 3GM/ab(a+b)$, G being the constant of gravitation. [M. T. 1926]

30. The surface of a uniform solid of mass M is an oblate spheroid of semi-axes $a(1-\epsilon)$, a, a, where ϵ is small; find, correct to the first power of ϵ, the gravitational potential of the body at an external point.

The body rotates slowly with constant angular velocity ω about its axis of symmetry $A'OA$, O being the centre of the spheroid, and a particle rests on the surface at a point P such that $A\widehat{O}P = \theta$. Find an approximate expression for the force exerted by the particle on the body, and shew that its direction is inclined to OP at an angle

$$\left[\frac{3\epsilon}{5} + \frac{a^3\omega^2}{2\gamma M}\right]\sin 2\theta,$$

approximately, where γ denotes the constant of gravitation. [M. T. 1933]

31. Two masses M are placed at distances c on opposite sides of the centre of a gravitating sphere of liquid of radius a and total mass M'. Shew that, neglecting powers of a/c above the third, the liquid is deformed into a prolate spheroid, the ratio of the minor axis to the major being

$$\left(1 - \frac{15}{2}\frac{M}{M'}\frac{a^3}{c^3}\right).$$

If two more particles of the same mass M are placed also at distances c from the centre on an axis at right angles to the line joining the former pair, shew that the surface becomes an oblate spheroid, the ratio of the axes being the same as in the previous case. [C. 1930]